THE AGING REVOLUTION

THE AGING REVOLUTION

THE HISTORY OF GERIATRIC HEALTH CARE AND WHAT REALLY MATTERS TO OLDER ADULTS

MICHAEL J. DOWLING
CHARLES KENNEY
MARIA TORROELLA CARNEY, MD, MACP

Skyhorse Publishing

Skyhorse Publishing books may be purchased in bulk at special discounts for sales promotion, corporate gifts, fund-raising, or educational purposes. Special editions can also be created to specifications. For details, contact the Special Sales Department, Skyhorse Publishing, 307 West 36th Street, 11th Floor, New York, NY 10018 or info@skyhorsepublishing.com.

Skyhorse® and Skyhorse Publishing® are registered trademarks of Skyhorse Publishing, Inc.®, a Delaware corporation.

Visit our website at www.skyhorsepublishing.com.

Please follow our publisher Tony Lyons on Instagram @tonylyonsisuncertain

Please follow Maria Torroella Carney on X @mcarneymd

10 9 8 7 6 5 4 3 2 1

Library of Congress Cataloging-in-Publication Data on file.

Cover design by Kai Texel
Cover artwork by Getty Images

Print ISBN: 978-1-5107-7882-5
Ebook ISBN: 978-1-5107-7923-5

Printed in the United States of America

TABLE OF CONTENTS

DISCLAIMER:

The names of all patients in this book have been changed to protect their privacy.

AUTHOR'S NOTE

In the fall of 1967, I left my home in the village of Knockaderry, Ireland, to begin studies at University College Cork, sixty miles away. It was a half day's drive on the meandering byways of rural Ireland, and I was fortunate to hitch a ride on a milk truck for the entire route. I had never before visited or even seen a college, and did not know anyone who had been to college, save for the local priest and schoolmaster.

I knew only one thing for certain: in college, I would get an education. To me, this was a most precious thing. My undergraduate years at University College Cork as well as graduate studies at Fordham and Columbia universities in New York changed the course of my life and eventually brought me to the world of health care, where research, study, and learning lie within the cellular structure of the profession.

This is a book about the most basic intellectual instinct—the pursuit of knowledge. Throughout the broad health care universe, all of us learn and all of us teach every day. Medicine is a complex business, a new

Rubik's Cube every day. It takes a special breed of person to accept the ongoing challenges.

In this book we write not only about breakthrough ideas, but also about the remarkable medical professionals who brought those ideas to life. The men and women in this book have followed somewhat similar pathways: recognizing a serious problem related to aging adults, experimenting with possible solutions, conducting the often-grinding, peer-reviewed research, and working to scale a new program or policy. Our book is a story of progress in how older people are cared for, while at the same time, it is a story about the fascinating people who have nurtured that progress over time.

As president and CEO of one of the nation's largest health systems, I consider this book an essential element of our responsibility to teach and share knowledge. My coauthors and I are honored to recount the stories of these thinkers, researchers, innovators, and healers, and to celebrate their achievements.

Michael J. Dowling
New York City
February 1, 2024

PREFACE

Suman Thakurdesai, my grandmother—or in Marathi, my "Aji"—was born and raised in Bombay, India. She worked as a mail sorter in her local post office, and more than seventy-five years ago, she met and fell in love with a mail carrier there—my grandfather. I can still hear her stories of their love blossoming as she handed him the mail for his daily deliveries.

I mostly remember my Aji doing what she loved best—working hard from morning until night, chopping vegetables or cilantro, and shredding the coconut on the kitchen floor as she prepared the next family meal.

As my Aji grew older, she started having back pain and was found to have multiple compression fractures in her spine due to severe osteoporotic disease. In pain, she was unable to do the one thing she loved most—preparing our family meals. Unable to do that work, she grew depressed, lost her appetite, and eventually, her will to live.

One day I received the phone call I was dreading. My mother—sobbing through the phone—told me my Aji was dying. I booked a ticket and got to her bedside as soon as I could. When I did, I found the family gathered, candles lit, curtains drawn, hymns being sung, and my Aji lying in bed, ready to die.

I greeted her as I have always done, by touching her feet in a customary Indian gesture of respect. I whispered to her that I was there. She acknowledged me briefly before falling into a deep sleep. Relieved that she had survived, I slipped into old clinical habits and asked my family about the details: When was her last doctor's visit? What meds was she on for pain? Anything for her depression? Physical therapy? Turns out—nothing. For a broken back, my Aji was on a low dose of Tylenol. No wonder she was losing her will to live.

So, I went to see her doctor. I asked him about pain control, her depression, physical therapy. I still remember his response vividly.

"You know that she's very old. She can't tolerate more treatment. You are asking too much. She will soon be no more."

I stared at him in disbelief. His diagnosis and treatment plan were steeped in ageism and overt bias. He had failed, fundamentally, to appreciate what really mattered to my Aji. She was not looking for heroics. She wanted to cut the vegetables again, free of excruciating back pain.

And so, for a short period, I became her doctor. I titrated her pain medicines, started her on an antidepressant, and arranged for in-home physical therapy. I accepted that if this was her time, it would be her time, but I also wanted to give her the best possible chance to achieve what mattered to her. Within two weeks, my Aji was sitting in her favorite place to watch the sun set, having just climbed two flights of stairs to get there.

My grandmother lived for another ten years beyond this moment in time. She was there for the births of all three of my children; cooked hundreds of additional family feasts; and told me many more stories of the love she shared with my grandfather.

Caring for our elders is perhaps the best exemplification of both what it means to be a healer, and why so many of us entered the health professions. No group of individuals has touched more lives, imparted more wisdom, and shared more love than our older adults. Delivering the best care to this unique population is at a critical juncture.

We should pause for a moment and celebrate the significant progress we have made over the past century in improving overall longevity. But we must also prepare ourselves for this landmark demographic change. Adults living longer means they can attend more weddings, more graduations, and more births of grandchildren and great-grandchildren. At the same time, this increase in longevity will stress our health systems. Nearly 95 percent of older adults have at least one chronic condition. Nearly 80 percent have two or more. Additionally, older adults suffer harm at the hands of health care at rates that are higher than other populations.

Fortunately, as with many other areas of great need in health care, there is real promise that things are improving. That's what this excellent new book is about—the people and ideas that are transforming how we care for older adults. This collection of stories and profiles charts a path through a variety of complementary efforts to improve care for older people. Much of the modern field of geriatric medicine was developed and refined by the individuals described in these pages. And this book reads as both a history of the field and a beacon of light to guide the future of the profession. These accounts not only relate the significant developments in geriatric care, they also describe a translatable approach that can benefit every other part of our health system.

This approach has many important elements—curious and dogged investigation, a refusal to accept the status quo, a dual belief in the power of both data and stories to change minds and improve practice. At the heart of all these stories is empathy. We may not be able to walk in the shoes of the older adults we treat, but we can and we must listen to them. We can and we must slow down, center the whole person in front of us, and really, truly listen. Even if we speak a different language, we must still listen—to the translator, to their family, to all the loved ones they have trusted to ensure their health and happiness. Only if we really listen to what's important to them will we be able to sustain and further the improvements described in this book.

I'm grateful to Michael, Charlie, and Maria for including a chapter on the work the Institute for Healthcare Improvement (IHI) and its partners at The John A. Hartford Foundation, the American Hospital Association, and the Catholic Healthcare Association are doing to propel an Age-Friendly Health Systems movement. This movement aims to enshrine a simple set of key evidence-based interventions for caring for older adults into everyday practice, and eventually, into policy and payment. Taken together we call these interventions the "4Ms"—Mentation (e.g., depression, dementia, delirium), Mobility (e.g., falls prevention), Medication (e.g., polypharmacy and reconciliation), and understanding and acting upon What Matters to older adults. Readers will surely notice each of these 4Ms in the other chapters and indeed our work in Age-Friendly Health Systems is built on the work described in those chapters. The first three Ms are critical to effective care, but it is the fourth—finding out what really matters to the patient—that has proven revolutionary.

I've come to believe that trust is the essential, and often missing, ingredient to excellent health care. Our failure to deliver the care that really matters to older adults and their families is a violation of a sacred trust. It might often go unsaid, but we go through our lives, making our own unique contributions to collective knowledge and collective progress, believing that when we reach the last stages of our lives, our wishes about the care we want will be heeded and we will be taken care of. We not only place trust in our families to carry out this important duty, we also place trust in the institutions of our societies.

Robert Butler and the many other leaders profiled in this book saw, earlier than most, that we were failing at this responsibility. Their work to correct this abdication of duty has changed millions of lives for the better. Since nearly all of us will grow old, all of us should be grateful to them. And all of us in health care need to learn from their example. This book is an important step in that learning journey.

My Aji should have never lost her will to live. She should have never had to rely on her grandson becoming a physician and being able to come to her bedside. Maybe if her health system had trusted her enough to ask her what mattered to her, and trusted that she still had so much more to give to and share with her family, she wouldn't have suffered so much.

This book should be required reading for all health care professionals. The lessons it holds and the warnings it carries are signals to our systems and societies of a lasting need to change and improve. I hope you will be as moved and inspired as I was to read it. And to all of you who have devoted your lives to caring for our elders—people like my Aji—I thank you.

Kedar Mate, MD
President and CEO
Institute for Healthcare Improvement

Chapter One

THE MAN WHO SAW OLD ANEW

"Perhaps more than anyone else, [Dr. Robert Butler]
is responsible for establishing geriatrics as a formally
recognized medical discipline in the United States."
—*The New York Times*, March 9, 1997

"A PERIOD OF QUIET DESPAIR . . . AND MUTED RAGE."

What was it, exactly, that drove Dr. Robert Butler to try and change American medicine in such a profound way? What fired Butler's decades-long passion to do nothing short of reimagine both the way the medical community cared for older adults and the manner in which doctors were educated and trained to provide that care?

When we search for clues to Butler's inspiration, it is necessary to travel back to the 1930s when Butler's parents split before his first birthday. Born in New York

City, he was raised in Vineland, New Jersey, by his grandparents until his grandfather, a chicken farmer, died when Butler was seven years old. The year was 1934, and the Great Depression came crashing down around Butler and his grandmother. They lost the chicken farm, and as Butler later recalled, "She and I were soon on relief, eating government surplus food out of cans... Grandmother found work in a sewing room run by the (federal) Works Progress Administration, and I sold newspapers and fixed bicycles for ten cents an hour. We moved into a hotel. When I was eleven, it burned to the ground with all our possessions."[1]

They started over, powered by little more than his grandmother's resolve. Through the hard times, the memory imprinted on Butler's mind was the adaptability of his grandmother. She was strong, optimistic, and did whatever was necessary to care for and protect her grandson. "What I remember even more than the hardships of those years was my grandmother's triumphant spirit and determination," Butler wrote. "It was my grandmother in the years that followed who showed me the strength and endurance of the elderly."

Butler excelled in high school, served in the Merchant Marines, and proceeded to Columbia University for undergraduate studies and medical school, earning his MD degree in 1953 at age twenty-six. During training in psychiatry at St. Luke's-Roosevelt Hospital Center in New York City, Butler was drawn to treating older people even as he realized that he had learned little in school about their particular medical and emotional needs.

"It began to hit me," Butler wrote later. "A lot of patients I was seeing clearly had psychological, behavioral, and social issues, not just medical ones. It also occurred to me that we didn't know anything about aging. So I thought there's something about old age that I've got to know."[2]

Butler joined a team at the National Institute of Mental Health (NIMH) in Washington, DC, exploring various aspects of healthy aging. He was captivated by the subject of growing old, not only from a medical point of view, but also from a social and cultural perspective. As he deepened his study of the aging experience, he grew ever more impatient with prevailing attitudes about older adults among his medical colleagues, many of whom preferred avoiding elderly patients.

By the late 1960s, Butler was fed up with what he saw as egregiously bad treatment of older people. "The medical profession and other health personnel share the culture's negative attitudes toward the old," Butler wrote. "In the medical context, this can take the form of active avoidance and dislike or a less-obvious pattern of paternalism and infantilism, pained tolerance, or caretaking rather than aggressive positive forms of treatment."

In 1969, Butler wrote the book *Ageism: Another Form of Bigotry*. His research and writing gained attention not only in medical circles, but also in Washington, DC, where President Ford chose Butler as the leader of the newly created National Institute on Aging (NIA) in 1975. Butler was a charismatic individual—tall, thin, and handsome, with an appealing combination of warmth and optimism. While he came to be recognized as one of the nation's leading experts in geriatric medicine, he never actually trained in the specialty for the simple reason that at the time of his medical training there was no specialty in geriatrics—not in America, at least.

At the time Butler was selected as leader of the NIA, he published his *magnum opus*, a 1975 book with the dark title of *Why Survive? Being Old in America.* The book, with its combination of tough judgments and constructive proposals, had an impact. For one thing, it was awarded the Pulitzer Prize in non-fiction, a rare achievement for a nearly five-hundred-page tome about aging and medicine. More importantly, the book struck a chord with some other physicians who felt as Butler did—that older patients were often neglected, and that medical schools and training programs fell short of preparing doctors to care for their complex needs.

In *Why Survive?* Butler wrote with the subtlety of a jackhammer. For many aging adults, he observed, old age was "a period of quiet despair, deprivation . . . and muted rage." Getting older, he wrote, was "often a tragedy . . . in a society which is extremely harsh to live in when one is old. The tragedy of old age is not the fact that each of us must grow old and die, but that the process of doing so has been made unnecessarily and at times excruciatingly painful, humiliating, debilitating, and isolating through insensitivity, ignorance, and poverty." This courtly physician with "a gentle countenance," as the *New York Times* put it, pulled no punches. How grim was Butler's portrait of aging in the 1970s? The *Times Book Review* noted that "Butler questions the value of long life for its own sake; modern medicine, he says, has ironically created 'a huge group of people for whom survival is possible but satisfaction in living elusive.' He proposes sweeping policy reforms to redefine and restructure the institutions responsible for" the care of the aged.[3] "We must ask ourselves," he wrote, "if we are willing to settle for mere survival, when so much more is possible."

The medical status quo in Butler's early days was that the "elderly" were seen as no different medically than other adult patients. Regardless of age or health status, adults were cared for by primary care doctors, yet very few of these physicians had specialized training in the care of older people. This was dangerous, Butler argued. He called for establishing a new medical specialty—recognized as having the same stature and legitimacy as cardiology or oncology, for example—focused on the complex needs of adult patients. In *Why Survive?* Butler observed: "Doctors and health personnel are not trained to deal with [the] unique problems" of older people whose "medical conditions are not considered interesting to teaching institutions, and they are stereotyped as bothersome, cantankerous, and complaining patients."

TOO MUCH SUFFERING

Why do we start our book with Butler's view of the aging experience? Because he defined what he considered the nadir of care in the modern age for older patients. In a sense, our book is a follow-up to Butler's—nearly five decades later—during which time the landscape has changed markedly. We argue in this book that the experience of aging has improved by orders of magnitude since Butler. While life expectancy in the United States has declined in recent years because of the pandemic and drug overdose deaths fueled by the fentanyl crisis, it's still thirty years higher (76.4) than it was in 1900 thanks to cleaner drinking water, vaccinations, and other public health initiatives. Campaigns against tobacco and drunk driving have also played a role, as have preventive measures such as screenings and treatments for cancer and heart disease.

But there is no escaping the price exacted by longevity. In the twenty-first century, a longer life brings increasing medical complexities, including openings for opportunistic diseases that degrade the thirty trillion cells within the human body. What is the verdict on this extra three decades of life gained since 1900? For many, time brings health and joy, for others illness and suffering.

No one escapes this life without some adversity, yet we make the case that much suffering among older people can be mitigated. In fact, that is what the men and women whose stories we tell in these pages have done and continue to do—ease suffering among aging adults. We explore the life-altering and often life-*saving* advances since Butler nailed his manifesto to the front door of the house of medicine. We argue that the medical community in particular and our society more broadly now have the knowledge needed to ease much suffering. Beyond the scientific, technological, and financial elements that help reduce suffering, the most powerful force of all is the ethical imperative imbued within our culture to care for seniors.

Permitting aging adults to suffer unnecessarily is an offense against an elemental ethical sensibility. The impulse to care for and comfort the old seems to have been stamped on much of humanity's DNA. Asian cultures, including Japan and China, set the standard more than 2,500 years ago when Confucius wrote that "filial piety" lay at "the root of humanity." Confucius defined filial piety as an obligation to provide the love, respect, obedience, companionship, and material support aging parents need throughout their later years. This aspect of Chinese culture has remained so durable that it is estimated that 70 percent of aging Chinese parents live with their adult children. In America, the power of family obligations has 40 million people caring for an aging relative at home. Americans long ago decided that supporting older adults is a collective responsibility—thus, the enactment of Social Security in 1936 and Medicare in 1965.

Butler was determined to shake the medical establishment out of its somnolence and jump-start a new movement to convince the medical world that older people should be treated differently in light of their varied and complex medical needs. At the same time, he sought to persuade the medical establishment that doctors needed to train differently to gain the competence needed to care for older patients. Butler's opportunity to ignite the new movement stemmed from his role at the National Institute on Aging (NIA), from his book, and from his role as head of the first Department of Geriatrics in the United States founded at The Mount Sinai School of Medicine in New York in 1982. Starting the following year, the Mount Sinai geriatric fellowship trained doctors who would go on to play leading roles throughout the country and, in the process, change how older people were treated.

"ADDING LIFE TO YEARS"

Among the foundational elements that reduce suffering is the sheer genius of medical researchers and practitioners. The new reality of aging is built in part upon a foundation of scientific advances in the treatment of heart disease and cancer, the leading causes of disability and death among older people.

The most significant advances for heart health come from two discoveries in particular: medications that stabilize blood pressure and reduce cholesterol. British physician James Black developed beta-blockers in the 1960s–70s, while Japanese biochemist Akira Endo made the original discovery that produced a new class of cholesterol medications.[4] Cardiac medications, devices, and lifestyle changes combined to reduce the age-adjusted mortality rate for cardiac disease by a remarkable 70 percent from 1968 to 2017.[5]

The second-deadliest threat to aging bodies is cancer, a disease for which the leading risk factor is age. Eighty-eight percent of people diagnosed with cancer in the United States are fifty years of age or older and 57 percent are sixty-five or older. Since 1991, there has been a 33 percent decline in the death rate from cancer—"an estimated 3.8 million deaths averted."[6] The overall cancer survival rate has climbed from 49 percent in the mid-1970s to 68 percent in 2023, while survival for breast cancer has reached 91 percent. Cancers of the thyroid, prostate, testicles, and melanoma have reached survival rates well over 90 percent.[7]

While cardiologists and oncologists have improved life for aging adults, so too have orthopedic surgeons with their ability to get patients back up and playing sports, hiking, running, and romping with grandchildren. Even with these advances, suffering among seniors remained common, in part because medical specialties such as cardiology and oncology typically existed (and often continue to exist) within disease-based siloes. Most older people, however, suffer from multiple maladies that require an un-siloed, team-based approach and the skills that come with geriatric medicine.

Among Butler's earliest and most influential colleagues in the early 1980s were Drs. Christine Cassel and Diane Meier, both of whom had started their training at the Veterans Administration (VA) Hospital in Oregon before Butler recruited them to join him at Mount Sinai. At the time, geriatrics in the United States was such a laggard that Cassel had little choice but to travel to England to complete her second year of training. "There was nobody here equipped to teach it," Cassel told us. "All the textbooks came from England and Scotland. There just wasn't any geriatric medicine in the United States." In sharp contrast to the United States, geriatrics is the second largest medical specialty in the UK and the top choice in UK residency programs.[8]

"YOU MEAN TO TELL ME WE'VE BEEN NEGLECTING OUR OLDER PATIENTS?"

Butler, Cassel, and Meier were of like mind. During their training, they had been disturbed by a callous attitude among some physicians toward older people. [Coauthor Maria Torroella Carney, MD, was trained and mentored by Drs. Butler, Cassel, and Meier at Mt. Sinai].

"I couldn't *not* see the suffering in the hospital," said Meier, who went on to lead the Mount Sinai palliative care program for more than three decades. "It was clear that no one was responsible for the suffering. It wasn't anybody's job to notice it, to assess it, to try to ameliorate it."

As Samuel Shem noted in his 1978 book *The House of God*, sometimes, when a frail elderly person was admitted to the hospital, some staff members would refer to the individual as a "GOMER," an acronym for *get out of my emergency room*. Many doctors at the time felt powerless to improve the health of frail elders who were often immobile and depressed. This attitude was born of ignorance. How could doctors going through a general internal medical training know about the special needs of old people when it was the gaping hole in the curriculum?

Butler and colleagues saw that, beyond the miracles of science in heart disease and cancer, another foundational element was needed—the expansion of geriatric medicine throughout the United States. Establishing new specialties in the field of medicine, however, is as difficult as it is rare. For Butler and colleagues, it was challenging in a hostile environment during the 1980s. Doctors in other clinical departments, particularly internal medicine, thought the idea of a geriatric specialty made no sense. "We are already taking care of older people," was the refrain. Some doctors saw the new initiative as insulting—"you mean to tell us that you think we have been neglecting our older patients?"

Butler and colleagues, however, saw that doctors knew very little about treating conditions prevalent among seniors, including frailty, falls, incontinence, osteoporosis, cognitive impairment, delirium, and polypharmacy, which is the mash-up of side effects from over-prescribing. These *geriatric syndromes* were responsible for suffering among millions of older people and their loved ones and yet the medical status quo of the 1980s accepted such suffering as an inevitable part of aging.

"The job of a geriatrician is to keep the patient going," Cassel told us. It is the *un*-siloed medical specialty. Good work by geriatric team members, including pharmacists, physical therapists, and social workers, controls multiple conditions in patients with heart failure, diabetes, osteoporosis, and hypertension. The idea in geriatrics, says Dr. Mary Tinetti from the Yale School of Medicine, is not so much to cure as it is to *manage* diseases and chronic conditions—often many different ailments in each patient. As Tinetti puts it: "The combination of medical, rehabilitative, palliative care, along with social supports and services" can ease suffering "and make most people live much better lives, even when they have all of those complex chronic conditions."

Dr. Barbara Paris was among the fellows at Mount Sinai who practiced under Butler during the early 1980s. She defines a good geriatrician as "someone who comprehensively takes care of an older patient" from about age sixty-five through the end of life. In the early stages of aging, when people are relatively healthy, this means focusing on steps to maintain health, including mammograms, colonoscopies, blood pressure stability, immunizations, and other preventive measures and encouraging a healthy diet, regular exercise, and the preventive and healing power of socialization. As people age and develop more complex problems, a geriatrician guides the patient on their journey through the medical maze, identifying specialists based on the patient's specific needs—in some cases, this can be as many as half a dozen—and coordinating care from home to office to hospital to rehab and back home again.

"You have to be able to provide care and know when you need a consultant," says Paris, "and you have to communicate with the consultant, and you have to still be the center of that patient's care. You must know what that cardiologist said or did. You have to be in the center of that." This sounds like common sense, but in a fragmented system it is an ongoing challenge.

Butler, Cassel, Meier, and others in the field rejected the status-quo treatment method of waiting until someone is sick and then bringing them in for an appointment. They believed in preventive measures, and the old-fashioned practice of house calls. They were visiting patients at home years before Medicare was willing to pay doctors for such visits.

"Home care," Butler wrote, "should be brought to the chronically ill, to those recuperating from hospitalization, to those who may be acutely ill but can be treated outside a hospital, and finally to all those older people who simply need some help from time to time to fill in the gaps left by declining functions and the loss of friends and relatives."[9]

As anyone working in medicine knows, doctors, nurses, and administrators in health care systems tend to be protective of their turf. Conflicts over control of divisions and departments are common, and so it was in the early days of geriatrics at Mount Sinai. When Butler established a requirement for internal medicine residents to spend some time rotating on to geriatric service, many doctors in training resisted.

"There were nasty comments," recalls Cassel. "Sometimes we were treated with indifference and even hostility by colleagues whose attitude was 'Why do we need you? And who with any ability would be a geriatrician anyway? Why should we listen to you?'" While Meier was participating in the geriatric training program, she recalls some people asking her why she would train in geriatrics.

"You're smart," she was told. "You could be a cardiologist."

"Geriatrics was seen as a consolation prize," says Cassel. "Of course, the opposite is true. It's far more intellectually challenging than any other subspecialty of medicine because of the number of forces, contingencies, and realities that interact on your patient that you have to take into account to provide good care. Internists saw it as kind of threatening, and they thought that all this stuff with the interdisciplinary teams was kind of a waste of time. It was seen as a soft science and therefore undervalued."

THANK YOU, MARTHA STEWART!

A serendipitous development helped elevate the status of geriatric medicine. Like other major New York hospitals, Mount Sinai was supported by many wealthy donors with aging parents or who were themselves aging. Some of these people with ready access to an array of physicians "were finding that all of their gold-ribbon specialists that make up New York health care weren't really helping," recalls Cassel, because they would focus on one of the patient's conditions—cardiology, for example—and ignore the three or four other conditions with which the patient was afflicted.

When these donors and trustees learned about the effectiveness of the geriatric approach, Cassel recalls, "They would call Butler when he was chair, or they would call me when I was chair and say 'Can you help me with my mom?' And then we would go and make a house call, which blew their mind. We would have a team brought in to make a full assessment of the issues—very often, so-called non-medical geriatric syndromes—and we would coordinate her care and make her better. That sold it right there." Among the donors was Martha Stewart, whose aging mother desperately needed comprehensive care. Stewart was so impressed with the way the team cared for her mother that she provided funding for a geriatric outpatient clinic, among the first in the country.

CLINICAL SUPERPOWER: THE RIGHT MEDICATION AT THE RIGHT DOSE

Butler envisioned a future where new treatments and drugs would ease suffering, and it must be said that progress within the pharmaceutical industry is nothing short of a modern miracle. Arguments will always swirl around the costs of these drugs, but that is a debate for another day. The fact is that throughout the world brilliant scientists are pursuing arduous, original research in biochemistry, physics, chemistry, biology, physiology, biophysics, neurology, gene therapy, and multiple other disciplines. Their knowledge leads to drugs that have improved the length and quality of life for billions of people. We are in awe of their work. Yet, many of these new medications have not been tested in populations who are older or who have multiple conditions and decreased organ function contributing to an unknown safety profile for older adults.

At the same time, we are well aware of the complexities involved in prescribing these medicines to aging adults with multiple medical conditions, and then managing the course of treatments. That is why the art and science of prescribing drugs has become a geriatrician's superpower. Effectively prescribing medications is not the type of miracle-work that makes front-page news, but it is, nonetheless, one of the most important things that a doctor can do for an older patient. Medication reconciliation is an often-complex process that involves identifying what medications have been prescribed for the patient, determining which drugs the patient is actually taking, calculating the various interactions between and among the drugs, often un-prescribing some meds, and finally, settling on the right medications in the right doses that work for the patient.

There is an unfortunate irony at work throughout the care system for older people: Medications intended to solve problems can cause new problems or exacerbate existing ones. One powerful drug in the weakened system of an older body has a particular effect. What happens when there is a second drug added? Or a third? Or a tenth? Medication that may have worked for decades may have a different effect as the aging physiology of kidneys, liver, and protein status evolves. In the United States about half of all people aged sixty-five and older are currently taking *four, five or more* prescriptions with complex interactions and side effects. Older adults tend to have multiple physicians, who each prescribe a medication to treat a particular disease. The problem is exacerbated when a patient is hospitalized and the care team is unaware of the person's medication history, perhaps due to a failure to consult with the patient's primary care physician.

We spoke with Dr. Kristofer Smith, chief medical officer at Optum at Home, our former colleague at Northwell Health, who encountered many patients on anywhere from seven to ten medications. Many other patients are on more than twenty medications.

"The challenge is disentangling why they're on some of these medications," Smith explains. "Some of them were started for a good reason a long time ago and no one reevaluated since."

One particular patient, typical in many respects, returned to the hospital soon after Dr. Smith and his team had treated him. "I looked at my team and I said, 'The only way we're ever going to figure out why he keeps getting high potassium is to go to his house.' While he was in the hospital, we got permission from his family to go to the home (with his family), and *lo and behold*, he had six different doses and types of blood pressure medications that all increase your potassium. We left with two bags full of medications he should never take again. It's not that uncommon, but it's hard to solve because if you're not going to the home and helping the family tease through what should be taken out of the home, they don't know."

Another patient brought in for evaluation at a Northwell geriatric practice (with Dr. Philip Solomon, a rising star in the profession) was believed, by her family, to be suffering from Alzheimer's disease. After a review of the timeline and history, it was revealed that one doctor had prescribed an anti-seizure medication with sedating side effects for a condition that had caused her severe facial pain. When she was in rehabilitation after hospitalization, she had been started on another, similar medication for facial pain—redundant medications, both with powerful sedating effects. The patient was also on blood pressure medicines and standing up made her dizzy. She presented to a geriatrician with dangerously low blood pressure, frail, drowsy, minimally verbal, weighing ninety-seven pounds.

After medication review and dosage adjustments, the patient returned a few months later rejuvenated, with better energy and awareness, and showing no signs of cognitive impairment. A physical therapist was sent to her home to help restore her strength and flexibility, and the family was astonished by her improvement.

Dr. Barbara Paris tells patients she has a rule: "You want to be my patient? Every visit, you have to bring in all your pill bottles. Every single one. What vitamins you take, what your cardiologist gave you, whatever you bought in the drugstore. If you don't bring those pills to me, don't come." In fact, medication reconciliation is recognized as so central to caring for older people that

Medicare now pays doctors more for visits that take place within the critical transition period of seven days after hospital discharge, which routinely include a medication review.

Geriatricians have long been fierce advocates for medication reconciliation and close supervision of transitions from one setting to another—hospital to home, home to hospital, hospital to rehab, rehab to home. Geriatricians have learned that older people often leave the hospital with a greater sense of confusion than when they arrived. Paris puts it this way: "The danger point for an older patient is when they enter the emergency department and leave the hospital and go home. Any time a patient touches any other part of the health care system, it's a danger point. And certainly, if they're hospitalized and then leave the hospital, that's very dangerous. So, the patient goes home with a list of medicines, and the list says furosemide [a diuretic] and before they came to the hospital, they were at home taking a different diuretic. The hospital substituted one diuretic for another, but the patient hasn't figured this out and they take both, and they faint because they're dehydrated and their electrolytes are off, and boom, they're back in the hospital. And they've been in a hospital bed for days and there's delirium, muscle weakness, they can't get up anymore. And nobody puts it all together for them."

THE OLD WILL OUTNUMBER THE YOUNG

Aging is not a static condition, but rather a process that unfolds gradually. The initial sign of getting older is perhaps that moment when the print on the page requires you to find stronger light, perhaps squint a little harder. The natural process of aging continues to degrade the strength and reliability of various parts of the human anatomy. The ability of cells to replicate diminishes, bones lose density, ligaments grow less elastic, and muscle mass decreases. The eyes see things called *floaters* that aren't there, while things that *are* there present as blurry impressionist brush strokes.

In 2022, there were 52 million Americans aged sixty-five or older, and that number has been growing by ten thousand *per day* and will continue to do so for years to come. Older people in America comprise a larger percentage of the total population than ever before. Perhaps most surprising is that by 2034, this nation—so long known for its youthful exuberance—will cross a threshold where *the old will outnumber the young*. Children and grandchildren of the baby boom generation will see levels of longevity previously considered unimaginable. An estimated half the babies born in 2020 will live for one hundred years or more, and experts predict that in decades to come it will not be uncommon for people to live beyond age 120.

With this expected longevity, the number of older Americans is projected to nearly double from 52 million in 2018 to 95 million by 2060, and this age group's share of the total population will rise from 16 percent to 23 percent. By any definition, these are mind-boggling numbers.[10] This population is not only growing in size, it is also becoming more racially and ethnically diverse. Between 2018 and 2060, the share of the older population that is non-Hispanic white is projected to drop from 77 percent to 55 percent, while the number of Black older adults in the United States will nearly triple and the number of Hispanic older adults will more than quintuple.[11]

Growth and diversity of the older adult population have significant implications for the US health care system. Demand for health care services will grow, the nature of the skills and services of the health care workforce will evolve, and the settings of care will change. Furthermore, the growing ethnic and racial diversity will create a need to address varied personal and family caregiver preferences, providing services with cultural sensitivity, and training the paid health care workforce in cultural competence.[12]

The oldest baby boomers, who famously warned against trusting anyone over age thirty, have celebrated (or, in some cases, endured) their seventy-fifth birthdays. The boomers carried the optimism of the American dream through the second half of the twentieth century and into the twenty-first, growing up through the Korean, Vietnam, and Cold Wars, the national trauma of assassinations, and the cultural and political upheaval of the 1960s. They cheered the space race, fueled the greatest economic boom in American history, and propelled advances in civil rights and equality for women. In 2022, the boomers were on the younger side of the aging scale as 23 million pre-boomers born between 1920 and 1945 reached their eighties, nineties, and beyond.

There is a tendency to lump older people into the same bucket, yet anyone searching for the "typical" senior citizen in the United States will do so in vain. The range of mental and physical well-being has almost infinite variations: Ninety-year-olds running marathons, sixty-five-year-olds incapacitated by stroke. More telling than numerical age among older Americans is biological age. From a technical perspective, biological age is measured by various properties of an individual's DNA and chromosomes; by whether people suffer from physical or cognitive impairments that accelerate aging. While a large percentage of older Americans sustain their health through diet and exercise, many others are inactive, smoke cigarettes, and consume prodigious amounts of alcohol and processed foods. Others, in contrast, have biological ages well below their physical ages and continue to play robust and central roles in business, science, the arts, education, medicine, and government, powering the nation's economic engine while also serving as custodians of an immense share of the nation's wealth. They are consumers, teachers, guides, and mentors who impact every aspect of modern life—though their contributions are not always recognized in our youth-obsessed culture.

Dr. Donald Berwick, a boomer himself at age seventy-five, the former administrator of the US Centers for Medicare & Medicaid Services (CMS) under President Obama, has developed a general theory that there are three broad categories into which most older people fall. "I think a third of people my age feel like I do, which is extremely healthy, vibrant," he says. "There's a wonderful quote from Gertrude Stein I read once. She said, 'We are always the same age inside.' Pick your age—I think mine's something around, I don't know, 15 or 16!" These are people who are generally healthy and independent, successfully managing various conditions and diseases.

Another third of older Americans, he says, are retired and enjoying time for leisure and family, and happily pitching in to help with and dote upon grandchildren. These are people who are enjoying the fruits of longevity.

Advances in medicine, nutrition, public health, and much more have enabled millions of aging adults to enjoy good health and a better quality of life than any aging generation in history. Whole industries have tailored their offerings to serve the elderly with the result that older consumers have choices in goods and services that would have been unthinkable a generation ago. While it is true that too many seniors struggle on Social

Security alone, the fact is, according to the *Wall Street Journal*, "Americans' increasing life spans have disproportionately increased the elderly's considerable wealth advantage. They've had more time to save and invest because of advances in medical science during their lifetimes." Many older people have disposable income and an appetite to enjoy life through travel, sports, and an array of leisure activities. The housing industry alone provides more options than ever with assisted living facilities and over-fifty-five communities.

Technological innovations helpful to aging adults come at a rapid pace. In Japan, thousands of sensors in a community track the whereabouts of older people with dementia to protect them from getting lost. Some caregivers are using virtual reality to enable patients to relive experiences from earlier, more joyful times in life. In Ireland, a depressed patient was brought to life by using virtual reality to "walk along the virtual Cliffs of Moher in western Ireland, just as he'd done with his wife several years earlier."[13] In the United States, the rapid adoption of technology to care for older people remotely took off during the pandemic. The best example was the sudden and widespread use of remote visits with patients via Zoom.

Another promising trend is the evolution from caring for patients in the hospital to doing so in ambulatory care centers and at home. In the future, the great majority of the care provided will be in settings other than hospitals, most notably the home, while hospitals primarily care for the sickest of patients. More and more hospitals will be seen as a link in the health care chain rather than as central hubs around which everything else revolves. Hospitals are great places to be if you have to be there, but not so great if you don't.

Unfortunately, what Berwick sees as the final third of the aging population are people with multiple medical concerns, lower incomes, and too much suffering. His off-the-cuff description jibes with more academic categories of older people—"young-old (ages sixty-five to eighty-four)" where many people enjoy a sense of emotional and physical well-being, then those age eighty-six to one hundred, many of whom have multiple complex chronic conditions. The third category also includes those over one hundred who often need a great deal of help.[14]

Many if not most older people in the United States enjoy good health or, at least, a quite acceptable level of health. Sixty-one million people are on Medicare yet only about 5 percent of that number account for nearly half of all Medicare spending.[15] These are the frail elders most in need of help; this group has the highest risk of falling, needing a caregiver, and being hospitalized.

In all, the suffering cohort totals about 2+ million people or about 1 percent of the US adult population. Our society is all too familiar with the concept of the wealthiest 1 percent of Americans. We contend in these pages that it is at least as important for our culture to become familiar with the *other* 1 percent—the older people who need our help.

"AMAZING THINGS..."

Robert Butler worked until three days before his death from leukemia in July 2010. He was eighty-two and had served as chair of the Department of Geriatrics and Adult Development at Mount Sinai until 1995. The *New York Times* obituary quoted Dr. David B. Reuben, chief of geriatrics at UCLA, saying that Butler "really put geriatrics on the map."[16] Butler's biographer, W. Andrew Achenbaum, wrote that Butler "helped to transform the study of aging from a marginal specialty into an intellectually vibrant field of inquiry . . . He initiated changes in the training of physicians and other health professionals on how to care for the elderly . . . Butler gave people reason to question stereotypes that demeaned late life . . ."[17]

Perhaps more than anyone else, Butler sparked the aging revolution in America. Since he and his like-minded cadre of clinicians began their work at Mount Sinai in 1982, geriatric capabilities have spread to every medical center throughout the country. In 1988, there were seventy-eight geriatric medicine programs in the U.S. That number had climbed to 162 by 2022.[18]

But one of the most discouraging trends in caring for older people is the dwindling number of doctors choosing geriatrics as a specialty.

"Demand for doctors specialising in geriatric medicine . . . is on the rise—and in many places the supply is rising to match. But not in America. Geriatrics is the least popular specialisation in internal medicine, and the country is facing a shortage . . . [T]he outlook in geriatric medicine is particularly alarming. According to the National Resident Matching Programme, which matches student doctors with hospitals, only 177 of the 411 spaces on fellowship programmes for geriatric medicine were filled in America [in 2023]. By contrast some fellowships, such as those in oncology and cardiovascular diseases, were entirely filled."[19]

Ultimately, for all the gloom in his book *Why Age?*, Butler proved to be an optimist. He wrote that "many of the ailments of the old are possibly preventable, probably retardable, and most certainly treatable . . . Major breakthroughs in the treatment of cancer, stroke, and heart disease are anticipated in the not-very-distant future. The time may soon come when old age is marked by a gentle and predictable decline rather than the dreadful, painful onslaughts of disease and chronic disability, which now haunt late life." Perhaps another indication of Butler's optimism for aging adults was his later best-selling book *Love and Sex After Sixty*, co-authored with his wife, Dr. Myrna I. Lewis.

The growth in geriatric and palliative medicine since Butler's days is testimony to his success but there are other indicators as well. From 2011 to 2021, a

University of Michigan study found older adults "experienced improvements in physical functioning, vision, and hearing, and through 2019, lower rates of dementia." A large provider of Medicare services found in a 2021 study that 72 percent of people over age 65 surveyed felt "younger than they are, with half saying they feel younger than" age 50.[20]

When persistent pain and discomfort are minimized or eliminated, and when aging bodies retain their strength and vigor, older people have the energy and optimism that makes for what matters most in life: deep emotional connections to our loved ones. Fundamental philosophical questions grow in significance as people age. Helene Fung, PhD, a psychologist at the Chinese University of Hong Kong, has written that "with age, people perceive future time left in life as increasingly limited. This sense of limited future time motivates older people to prioritize goals that aim at deriving emotional meaning from life."[21]

We trace the arc of improvement from Butler's days when specialties in geriatrics and palliative care did not exist in America. By 2023, these specialties could be found at virtually every health system in the United States and internationally. Geriatric syndromes like delirium and falls are being addressed routinely. Transitions of care and medication reconciliation are being prioritized. Prognoses are better understood and conversations about treatment options are happening routinely.

Dr. Eric Widera, Professor of Clinical Medicine, Division of Geriatrics, UCSF, and Director of Hospice & Palliative Care, San Francisco VA Medical Center, told us that some of the recent progress in geriatrics "would have been unheard of fifteen years ago. Completely unheard of. In geriatrics, we don't give ourselves enough credit and I think we've done pretty amazing things."

Since Butler's days American medicine has flourished in many areas. As we have seen, doctors have developed more-effective treatments for the deadliest diseases, created less-invasive surgical techniques, invented replacement parts for sustaining mobility, shifted the locus of care closer to patients' homes, improved management of common chronic conditions, and offered financial incentives for doctors to prevent illness in the first place. In its post-Butler incarnation, aging in America remains a challenge, but it is also, for many people, an experience where the chronic conditions that typically accompany old age are more manageable, where older people enjoy more options for work and professional development, education, leisure, travel, sports, and maintaining physical strength and mobility. For increasing numbers of Americans, life is healthier and richer in the experiences that matter most. These are the fruits of the aging revolution.

Health care is a self-critical industry. Scores of medical journals regularly publish articles outlining deficiencies in care at hospitals, physician groups, and rehab facilities and much of that self-criticism powers innovation. Too much becomes toxic, however, giving way to common cries that the US health care system is too expensive, unorganized, greedy, a mess. There is no other industry whose customer is every person in the country from birth to death. Any discussion or debate about the perceived faults of health care—expense, access, quality—must start with a reality check. How many families and their children are now able to celebrate as a result of the great work of health care providers and medical science? How many are flourishing because of successful surgery? How many are still alive because of care provided by hospitals and ambulatory facilities? Among adults, how many survived COVID-19 because of the vaccines and medical care they received?

Health care professionals should all be proud to be in this business and appreciate the special responsibility and obligation we have. This special role was evident during the height of the pandemic—but is also evident in the work done each and every day in America's medical facilities and in the homes of patients. In a fast-moving, loud and polarized world, normalcy can sometimes seem radical. Everyday work often goes unnoticed. The value of health care is not defined by its highest highs or lowest lows, but by the millions of moments in between when lives are improved, health is restored, and suffering mitigated.[22]

This is a story about ideas and innovations, but it is also a story about people—patients, of course, but also physicians and other clinicians, who have devoted their professional lives to making the revolution possible. In these pages we focus on a handful of people who have dedicated their professional lives to easing suffering and making a better life for aging adults. General readers are unlikely to recognize their names. They are not famous in a celebrity sort of way, but they are stars within the world of health care—admired for their determination and revered for the progress they have made in bettering life for aging adults.

INDEPENDENCE AT HOME

"Really smart people at the top of their game."
—Attorney Jim Pyles

MR. STANKOWSKI AND DR. DE JONGE

Dr. Eric De Jonge was in his second year of an internal medicine residency in 1992, when he first visited the East Baltimore home of Mr. Paul Stankowski, a retired Bethlehem Steel worker, and his wife Miriam. The Stankowskis lived in the Highlandtown neighborhood, a blue-collar area of narrow rowhouses constructed of grayish stone—functional homes with little ornamentation. It was a tough neighborhood of no-nonsense people, many of whom had worked at the sprawling Bethlehem Steel plant. Like other local residences, the Stankowskis' home was so small that when De Jonge would visit, the front door would bump into Mr. Stankowski's recliner in the sitting room. The recliner occupied a room about six-by-ten feet in size—this was where Mr. Stankowski, at age eighty, spent most of his time. Although the space was cramped, it included a small bookshelf alongside the recliner with photographs of his children and grandchildren.

Mr. Stankowski was a rugged man, stocky in build, with a ready sense of humor. He would listen carefully as De Jonge explained news of his latest tests. De Jonge diagnosed and treated Mr. Stankowski's chronic obstructive pulmonary disease, diabetes, heart failure, and metastatic lung cancer. At the time, the nation's hospitals were filled with patients like Mr. Stankowski, but a hospital was the last place he wanted to be. De Jonge's mission was to treat what was treatable, while managing the symptoms of other conditions that were not curable, enabling Mr. Stankowski to remain at home.

The avuncular town physician arriving at all hours with his trusty black bag is etched in history, yet by the 1980s and '90s, home visits by doctors were unusual. De Jonge's visits were part of a modest effort at the time to get back to days when the physician went to see the patient, rather than the other way around. To De Jonge and others of like mind, this one-to-one doctor-patient in-person treatment was more effective than the status-quo office appointment approach. Rare were the doctors drawn to this type of practice, but from his first days at Johns Hopkins Bayview Medical Center, De Jonge loved house calls.

Thirty-one years after that first visit to Mr. and Mrs. Stankowski in East Baltimore, De Jonge has vivid recollections of the experience that he still considers a privilege: an honor to be welcomed into the Stankowskis' home.

"Mrs. Stankowski was attentive, vigilant, and loving, and Mr. Stankowski was a salt-of-the-earth person—a crusty, no-nonsense guy."

De Jonge was barely twenty-eight years old at the time, yet he felt enveloped by the warmth of these people and their regard for him as *our doctor.* Mrs. Stankowski would offer De Jonge refreshments—tea, water, and snacks—upon his arrival. Typically, De Jonge's visits came in the early evening as the intensity of the workday began to ease. De Jonge would head home after the visits, "feeling like he touched human beings in a way that [he] could not do in the clinic."

The Stankowskis were comforted by his visits and the knowledge that the rest of the Johns Hopkins team members—nurses, social workers, and other clinicians— were available to solve any number of problems.

"It was the visits," recalls De Jonge, "but it was also knowing our team offered a lifeline to people to relieve pain, shortness of breath, make a diagnosis of lung cancer, talk openly about prognosis, and listen to his wishes about what he wanted and where he wanted to die. I think all that gave him some comfort. Being in the home allowed me to be more in a servant position as opposed to him coming to the office and getting a medicalized experience."

At Johns Hopkins, De Jonge also worked in the hospital, but his experience in the home of the Stankowskis and others set the course for his career.

"I found that I could do a better job of taking care of the most ill patients when I saw them in their homes. It was also an area of personal growth for me to be able to care for the most complicated, sick people with a medical, ethical, and emotional perspective, to manage them until the last day of life. It was both an intellectual and emotional challenge, and a profound, shared experience with the family and patient at the end of life."

In time, Mr. Stankowski succumbed to his lung cancer, after choosing hospice and comfort care at home. Not long after he died, De Jonge received a personal note from Mrs. Stankowski.

"She thanked me for having visited him at home," De Jonge recalled. "And for having made his life a little bit better."

HIPPOCRATES: GO TO THE PATIENT'S HOME

De Jonge's path toward medicine had begun while he was a student at Stanford.

"I wanted to see how I could have the biggest impact on justice in our country, social justice," he told us. "I majored in history and biology and was interested in public service." But during his undergraduate years De Jonge got to know Stanford President Donald Kennedy, a biologist and former head of the US Food & Drug Administration, and Kennedy suggested De Jonge pursue medicine, a field where he could care for patients while also working on policies to improve the broader health system. De Jonge followed Kennedy's advice and after graduation in 1986, he enrolled in the Yale School of Medicine, from which he graduated in 1991. During medical school, he took a year off and moved to Washington, DC, for an HIV/AIDS policy fellowship. De Jonge was a junior staffer on a National Academy of Sciences' report intended to evaluate a series of questions related to AIDS research conducted by the NIH: *How was the money being spent? What were the research priorities? What were the results?* De Jonge loved the work, but his pay was so low that he took a part-time evening job as a waiter at a Mexican restaurant to help cover his rent.

By the early 1990s, De Jonge had completed medical school, residency in Baltimore, a policy fellowship at Georgetown, and a geriatric medicine fellowship at Johns Hopkins. He was committed to both treat patients at home while also working to change how Medicare treated and paid for the care of patients. Medicare rarely paid doctors to care for patients in their homes, though that is exactly what De Jonge believed Medicare should do. In the 1990s, De Jonge had the good fortune to meet another physician who was also committed to caring for patients in their homes. Dr. George Taler, a modern pioneer in home care, had been among the founders in 1987 of the American Academy of Home Care Medicine, a small organization devoted to spreading home-based medical care for frail elders.

When we asked Taler when doctors first began to visit patients in their homes, he replied: "At the time of Hippocrates." But in the modern age, appointments in the doctor's office are the standard.

"Our current out-patient and clinic structure were designed to manage acute care concerns through rapid-fire doctor patient encounters," says Taler. "Unfortunately, our needs have changed dramatically over the past few decades. The default is that the most common out-patient conditions we encounter today result from lifestyle choices (obesity, smoking, alcohol and drug abuse, inadequate exercise, hypertension, diabetes, and hyperlipidemia), which are progenitors to the most common medical illnesses (heart, brain, lung and liver disease, and cancer). In addition to [medication] management, these factors take education, coaching, changing behaviors and engaging others in the family to help achieve the change. But, by design, we are given neither adequate time nor the support to address these challenges in this setting. The result is that only a small fraction of patients heed our advice and an even smaller fraction maintain adherence . . .

"The way to change all this," says Taler, "is to rethink how best to engage patients who would benefit from lifestyle interventions, but for the most complex, it is to care for people in their home. For the 'rising risk' cohort (patients with chronic conditions due to lifestyle and genetic factors, but who are not yet disabled by disease), the most efficient and effective approach is a combination of periodic office consultations to set goals and identify barriers to success, frequent telehealth check-ins, and group sessions with others who share your predicament. Physicians have the 'moral authority' and gravitas for the consults, but nurse practitioners and physician assistants have the knowledge, charm and prescriptive authority to manage the telehealth encounters and make adjustments to the regimen."

Medicine changed in the post-war era as care delivery grew more complex and the doctor's black bag became a symbol of a simpler past. Drs. Bruce Leff and John R. Burton have observed that during the first half of the twentieth century "a physician could carry the entire effective pharmacopoeia in his or her talismanic black bag: morphine, insulin, digitalis, and adrenaline." But medical breakthroughs, technology, and a trend in medical culture toward large institutions shifted care from home to hospital. These burgeoning academic medical centers were marvels of the modern world where lives were saved and health restored in ways previously unthinkable.

For all the grandeur of modern medicine, however, something was lost in the transition to institutional settings. Florence Nightingale understood this back in the nineteenth century when she established a nursing school to educate students on how to care for patients in a variety of settings, including in the home.

Fast-forward to the twenty-first century, where one of the most significant medical trends is providing care in the home, particularly for frail elders. Part of the explanation for this evolution toward home care lies with changes in how medical care is paid for—a subject we explore in greater depth later in the book. But another part of the explanation is the reality that suffering often hides in the home and needs to be treated there, yet Medicare was unwilling to pay much if anything for home visits. At the American Academy of Home Care Medicine, Taler and his colleagues sought to persuade Medicare to change its payment policies.

"We said, 'this is just not going to fly,'" Taler told us. "Who in their right mind would make a house call and get paid one-third" of the rate for an office visit. Thus began a sustained campaign by the Academy team to convince federal officials to increase payments and thus improve the quality of care for older patients. During 1989 and 1990, Taler and his colleagues met officials from the federal government and the American Medical Association, making a case that delivering primary care in the home to vulnerable patients improved quality.[23]

"We were able to get a premium for house calls and everybody was really surprised that a small group of people could actually pull this off," Taler told us. The result was an uptick in physicians as well as nurse practitioners conducting in-home visits.

De Jonge and Taler set out on their own in 1999 and established a new practice focused on home-based medical care, funded in part by the new Medicare reimbursement rates as well as gifts from philanthropic sources. They set up shop at MedStar Health, a large non-profit health system, and "built a team to serve the underserved part of Washington, DC, eight zip codes where we cared for a largely low-income, Black population," De Jonge told us. For the past twenty-five years, De Jonge, Taler, and colleagues have provided continuity of care to their patients with three elements—home visits, serving as attending physicians in the hospital, and providing 24/7 availability by phone.[24]

An essential element of home-based primary care is trust. "It has to be in-person to build that rapport," says Dr. Kristofer Smith of Optum. "These patients are really, really complicated." With many patients and families, says Smith, "if you send someone to the house when a family member calls, saying, 'Mom's not doing well,' they're your friend for life. Even if you could do some of it over the phone, there is a behavioral element of winning the trust battle with patients and families."

THE ACADEMY TEAM

The physicians who had gathered together in 1987 to form the American Academy of Home Care Medicine had succeeded in winning reimbursement increases from Medicare, but the organization was so underfunded that the office was located in space above a garage outside Baltimore. Though neither large nor wealthy, the group nonetheless had the ability to punch above its weight class. What it lacked in funds it more than made up for in the knowledge and passion of the doctors drawn to the group's mission. These were physicians who had devoted their professional lives to caring for aging adults, especially frail elders. The central theme that bound these doctors together was their determination to get out of the office and hospital and care for patients in their homes.

In the early 2000s, a number of Academy team members[25] began writing a proposal for a more robust program of home care medicine that they believed would be better for their patients. They believed that good medical practices and policies always had to come before cost, but they also knew that in the real world—the $766 billion world of Medicare, the second-most expensive government program in the United States (behind Social Security)—they had no choice but to pay attention to cost as well. The Academy team members believed that if they could demonstrate better care for their patients at the same or even lower cost to Medicare, there was a much greater chance of having government backing to spread home-based primary care throughout the nation.

The target population of the program included frail elders—mainly homebound—who suffered from many chronic conditions including dementia, heart failure, diabetes, sensory impairments, anxiety, and depression. These were patients who were unable to keep clinic appointments, had visited the emergency department, and/or had multiple hospitalizations in the prior year. This demographic typically consisted of the most complex and expensive patients in the Medicare program. In fact, the top 5 percent of Medicare users account for 43 percent of Medicare spending.[26]

Caring for these patients required a team—physicians, nurses, nurse practitioners, social workers, clinical pharmacists, and staff coordinators. It also required access to frequent in-person home visits by one or more members of the team, advanced monitoring technology, 24/7 live phone access, and the ability to manage care in the hospital when needed.[27] Delivering such an intensive level of in-home care was a daunting challenge. Fortunately, however, the US Department of Veterans Affairs (VA) had sprinted ahead of just about everyone in the field of home-based primary care and blazed a trail for others to follow.

VA TRAILBLAZERS

Initially, there was some skepticism in the broader

medical community about the VA work. To large legacy health systems, the VA can sometimes seem like a confusing bureaucracy. In fact, the Veterans Health Administration (VHA), the largest integrated health system in the United States, cares for more than 9 million veterans at more than one thousand ambulatory and inpatient facilities.

What is little-known about the VA, however, is that it also has set a standard for home-based primary care and age-friendly care. Many staff members at the VA contributed to this pioneering program, but among the leaders is Dr. Thomas Edes, a geriatrician who started in the VA system in 1984 in Columbia, Missouri. From 1991 until 2000, Edes led a home-based primary care program for patients in the Columbia area.

The personal touch of having a doctor right in your home was rare and beautiful to these patients, many of whom were eighty years old and above. One day in 1996, a senior administrator at the hospital told Edes that veterans and their families loved the program, as did physicians and other team members. Then came the *but*—the home-based primary care program was too expensive to maintain and cuts needed to be made.

Edes was stunned. Everything had been going so well. He lived within his budget almost to the penny. Everybody involved with the program was happy. As he tried to absorb this unsettling news, Edes posed a somewhat surprising query to his boss: "Before a final decision, will you give me two weeks to try and figure out an answer to one question?"

"What's the question?" the administrator asked.

Edes replied: "What is it going to cost us to *not* have this program?"

The administrator considered that for a moment and, to Edes's relief, said yes. Edes and his colleague Roger Langland, a social worker who played a central role in administering the program, got together and came up with a simple plan: They would retrieve the names of every veteran who had been in the VA home-based primary care program for the past two years and had also been in the traditional VA care program for the two years prior to that. And they would compare the two.

After a couple of weeks of analyzing the data, Edes returned to the hospital administrator's office to review his findings. Edes found that it was costing $6,000 per veteran per year for the home-based program (including all medication costs).

"These were among the sickest, most-complex patients who faced socioeconomic challenges and suffered from an array of conditions ranging from cognitive impairment to cancer, from heart disease to Parkinson's," he said. "These were patients who were frequent visitors to the emergency room and for in-patient hospital stays. They suffered from arthritis, hearing and visual impairment, falls, and urinary incontinence."

However, there was a surprising finding underneath

the data: The home-based medical care program was actually saving the VA $30,000 *per patient per year.* How? By keeping patients stable at home so they did not require expensive in-patient and rehab stays.[28]

Edes was suddenly quite popular. His colleagues throughout the VA system wanted to learn from him, as did physicians outside the system. The more deeply Edes delved into the issue, the more convinced he became that the VA could help the nation figure out how to improve care and reduce cost among millions of older Medicare recipients—particularly those in the highest-cost bracket.

In 2000, Edes moved to Washington to lead an expansion of VA home- and community-based services as well as palliative care. Two of his top goals in his new job were to establish a home-based primary care practice in every VA medical center in the country and to change Medicare by making home-based care an integral part of the program. In Washington, he joined forces with the American Academy of Home Care Medicine team in a lobbying initiative to persuade Congress to enact such a law. The Academy team, with Edes included, mapped out their idea for a new Medicare program, which they called *Independence at Home* (IAH). The Academy had the good fortune to find an experienced lawyer steeped in the ways of Washington who also happened to have a keen interest in improving health care for older people. Attorney Jim Pyles, who suggested the title Independence at Home, had spent the bulk of his career practicing before federal appeals courts, but when he met Edes, De Jonge, Taler, and the other team members, he saw an opportunity to be part of changing how the Medicare system worked. He had important experience at his law firm related to Medicare rules and reimbursement guidelines, and he knew the highways and byways of Capitol Hill.

Pyles told us that the Academy team members "were really smart people at the top of their games." For a fraction of his normal fee, Pyles signed on to help the Team sell their idea to Congress. The reputation Pyles and his firm had built through the years as straight shooters opened the doors to House and Senate members. Pyles took the detailed proposal outlined by the Team and wrote it into statutory language as a formal bill. Pyles would call ahead to make appointments and then accompany the doctors to the meetings.[29]

From 2007 to 2010, Taler's recollection is that the Academy team held about one hundred meetings on Capitol Hill to educate staff and House and Senate members about their IAH proposal. Early on, they met with Edward Markey, then representing Massachusetts in the House of Representatives (he was elected to the Senate in 2010), and with Senator Ron Wyden of Oregon. Markey and Wyden were the key supporters of the proposal in Congress.

"We would go into the meetings and the doctors were

just impressive," recalls Pyles. "They were very, very good. Their sincerity about what they were doing and the fact that they'd been doing it at a loss for years just because they knew that it was a service to patients and to health care. Totally devoted. And they had stories to tell. They wouldn't give the name of the patient, but they would tell how they had kept the patient who had previously had five ER visits . . . visit-free for a year or more."

De Jonge recalls the meetings followed a similar track. "We would say that 'you have a problem that we can help you solve. The problem is that Medicare costs are excessive, unnecessary, and increasing at a rate that is unsustainable. The good news is that the sickest, most-expensive patients who cost Medicare forty to fifty thousand dollars per patient per year includes a lot of unnecessary spending. If we do better care for them at home and keep them out of the hospital, ER, and nursing home, then we can have better outcomes for the patients. We can both improve the quality of the care and reduce the total Medicare costs.'"

THE MIRACLE OF BIPARTISAN SUPPORT!

The work that Tom Edes and his VA colleagues undertook—with high patient satisfaction and cost savings—added significant credibility to the Academy's proposal. The team would emphasize that the highest-cost patients, who represented 5 percent of all Medicare patients, accounted for nearly half of all Medicare costs.

"We made that argument over and over and over and over again," De Jonge recalls. He argued that potential savings for Medicare with a comprehensive home-based care program could reach $10 billion a year.[30]

The Academy team's pitch won over House and Senate members from both sides of the aisle. Early supporters, in addition to Markey and Wyden, included Senators Olympia Snowe and Susan Collins of Maine (both Republicans) and Edward Kennedy of Massachusetts (Democrat). Members of Congress recognized that the Academy team was bringing them something new and exciting. *You are saying you can provide high-quality care—with no rationing, no restricting care—and you will reduce Medicare costs by 10 percent?*

The Independence at Home Act (H.R. 2560) was formally introduced on May 21, 2009. The bill touted studies showing that "hospital utilization and emergency room visits for patients with multiple chronic conditions can be reduced and significant savings can be achieved through the use of interdisciplinary teams of health care professionals caring for patients in their places of residence." It was remarkable. De Jonge, Taler, and the rest of the Academy team members were seeking to change the delivery of care to frail elders in a profound way *and no one opposed them.*

But just five months later, on October, 29, 2009, President Obama introduced the Affordable Care Act and partisanship took over. Republicans were uniformly opposed to Obama's proposal overall, but were silent on the issue of home care. Pyles recalls that if Republicans "wanted to make an argument against the Affordable Care Act, they weren't going to start [with Independence at Home]. Because look, just think of the counter argument. *We are against a program that would provide better care to the elderly and infirm, and it saves Medicare money.* That's a tough thing to promise." While there was overwhelming opposition among Republicans to the ACA, there was similarly overwhelming support among Republicans for the Independence at Home proposal within the ACA.

In 2012, the Independence at Home Demonstration project launched in seventeen medical practices throughout the nation. Government publications are not often known for their clarity but in this case the Centers for Medicare and Medicaid Services described the program nicely as:

Home-based primary care [that] allows health care providers to spend more time with their patients, perform assessments in a patient's home environment, and assume greater accountability for all aspects of the patient's care. This focus on timely and appropriate care is designed to improve overall quality of care and quality of life for patients . . . while lowering health care costs by forestalling the need for care in institutional settings . . .

Three different home-based medical care practices led by Academy team members enrolled in the IAH demonstration project. These included Medstar Health in Washington, DC (Taler, DeJonge), the University of Pennsylvania in Philadelphia (Dr. Bruce Kinosian), and Virginia Commonwealth University in Norfolk (Dr. Peter Boling). Rather than create three different Independence at Home programs, however, the teams combined all three practices in a single group called the Mid-Atlantic Consortium. It performed well.

The Independence at Home model sought to create a deep sense of trust between patient and provider. When a person is first enrolled in the program, a physician goes to the home for a lengthy, in-person session to review medical history and ask an assortment of questions:

- *How are you managing during the day? Who is helping you when you need to get groceries?*
- *I see your home is very clean. How are you getting all this cleaning done?*
- *How much support do you have at home?*
- *What is the safety of your home? Are you at a fall risk?*

Physicians find that they are able to learn a great deal about patients during these sessions by seeing the home and asking about photographs prominently displayed, pictures that often seem to chronicle a life—marriage, children, perhaps grandchildren. People open up much more than they do in an often-hasty office visit. As they do, the tenor of the physician-patient relationship changes during the first visit. There is a warmth and intimacy to the encounter that benefits both the patient and medical team. We note with pride that our Independence at Home initiative at Northwell Health, led initially by Drs. Kristofer Smith and now Konstantinos Deligiannidis, was ranked as the top performer in the United States for the eight years of the program thus far.

"NOT YOUR GRANDFATHER'S HOUSE CALL."

Prior to the launch of Independence at Home, De Jonge initiated an effort to apply academic rigor to study the outcomes of home-based primary care. He began the research in 2010 and the resulting paper was published in 2014. His research applied statistical methods to create comparable cohorts of patients to test whether home care medicine could provide affordable, high-quality care. He already knew the answer—given that he had been practicing home-based primary care for eighteen years at this point—but the scientific community needed proof, so he and colleagues in DC set out to study the issue and publish their findings in a credible academic journal.

With data from Medicare, along with the guidance of an expert statistician who adjusted for comorbidities,

severity of illness, and disability among the patient population, De Jonge and colleagues set to work.[31] They
found that comprehensive, home-based primary care for
the most-ill elders worked as their front-line experience
had demonstrated through the years. In the paper, published in the *Journal of the American Geriatrics Society*,
De Jonge and his colleagues wrote that "patients who
received 24/7 intensive, home-based primary care had
total Medicare costs lowered by 17 percent over two
years." Most of the savings came from providing regular and urgent house calls, which prevented unnecessary
and expensive hospital visits.

The lesson of IAH is that a dedicated team that provides 24/7 live access and also follows patients across
multiple settings—home, emergency department, hospital, rehab—can achieve excellent quality results and
reduce Medicare costs by as much as 15 to 30 percent.[32]
De Jonge's paper and Edes's work at the VA arrived at
similar conclusions. The results were impressive and people noticed. Not only did doctors in not-for-profit health
systems notice, so too did investors at private equity
firms and the largest insurance companies. Perhaps the
combination of these three indicators served as a catalyst for an avalanche of private investment made in systems that care for sick, older people. (Overall, as previously noted, studies have shown that providing in-home
care stabilizes costs and often reduces the cost of care.
However, it should be noted that home care for some

highly complex patients can be even more expensive than in-patient hospital care.)

COULD HOSPITAL CARE BE MOVED TO THE HOME?

Dr. Bruce Leff was a fellow in geriatric medicine at Johns Hopkins Hospital in 1992 working on home-based primary care. Coincidentally, like De Jonge, Leff also trained early on at Johns Hopkins Bayview Medical Center in Baltimore. The two men shared a similar motivation: *to reduce suffering among older adults.*

Like De Jonge, Leff loved home-based primary care and found that treating older people at home was not only effective medicine, but also an "uplifting experience" for both patients and doctors. His patients were homebound due to various functional impairments and much of Leff's work was aimed at managing patients' conditions effectively enough so they would not have to be hospitalized.

"A lot of our patients did not want to go to the hospital," he told us, "because they had been there before and had terrible experiences by virtue of the hospital environment, not simply because they were sick."

Hospitals can be hazardous places, particularly for older people and for patients with dementia, while emergency rooms and hospital floors can be frightening and disorienting. Inpatient rooms often lack the calm, healing environment that elders need, considering that many have high rates of delirium, and are at risk of pressure sores and hospital-acquired infections. At the same time, it is also true that America's hospitals routinely provide miraculous care—saving lives, healing diseases, restoring health to millions of people with extraordinarily complex conditions. When a patient is in crisis nothing compares with the personnel, knowledge, experience, and technology of America's hospitals. A point often forgotten in discussions about home care is that, in many cases, it is thanks to a hospital stay and the work done by teams there that an older person is able to achieve a level of stability that allows them to transition back home.

In 1994, Leff began to explore the idea of hospital-at-home. Leff wondered whether there was a way to marry hospital and home care so that patients suffering an acute illness could receive hospital-level quality and intensity at home. Leff envisioned a health-care delivery system where primary care and hospital-level care could both be provided in the home thus, Leff says, turning hospitals of the future into "big ERs, ORs, and ICUs" with "everything else out in the home and community."

His idea has become a reality in some places. Many of the best practices of hospitals are being adapted within hospital-at-home programs that aim to honor the wishes of patients and family members "to be at home, but also to remedy hazards of hospitalization . . . As geriatricians, the main focus was to help our patients avoid the hazards of hospitalization, which for older adults are substantial: losing the ability to function, falling in the hospital, adverse drug events, developing delirium, the risk of those poor outcomes," says Leff. "Someone comes to the [Emergency Department], they need hospital-level care, let's say it's heart failure or pneumonia, they meet the medical criteria requiring hospital-levelcare. They need physicians, nurses, IV fluids, IV medicines, X-rays, blood tests, echocardiograms. You can do all that stuff at home."

The New York Times would later report Leff noticed "in the late 1980s while making house calls to home-bound patients [that some] . . . simply refused to go to a hospital. He understood why: he had seen the delirium, infections, and deconditioning that too often land older patients in nursing homes after hospitalization. Leff and his colleagues had an idea: What if patients could be hospitalized in their own beds?"[33] The *Times* reporter wrote that:

> ". . . under pressure to reduce costs while improving quality, a handful of hospital systems have embarked on an unusual experiment: they are taking the house call to the extreme, offering hospital-level treatment at home to patients . . . who in the past would have been routinely placed in a hospital room. And as awareness spreads of the dangers that hospitalization may pose, particularly to older adults, patients are enthusiastically seizing the opportunity . . ."

The *Times* reported that "the findings, published in the *Annals of Internal Medicine*, were promising. Offered the opportunity, most patients agreed to be treated at home. They were hospitalized for shorter periods, and their treatments cost less. They were less likely to develop delirium or to receive sedative medications, and no more likely to return to the emergency room or be readmitted."

While the results were encouraging, there remained a major obstacle: neither private insurance companies nor government health plans such as Medicare would pay the costs of Hospital at Home (other than a few experimental programs). Hospitals prosper financially with "heads in beds," as the old expression goes. Steering prospective patients away from the hospital was contrary to the financial benefit of the institution.

The COVID-19 pandemic changed that overnight. With hospitals pushed to the limit, constructing ICUs in hallways and employee cafeterias, something had to give and that something was payment. In November 2020, Medicare announced it would pay for Hospital at Home care at the same level it would pay for in-hospital care—a historic breakthrough. Prior to this Medicare policy, about twenty hospitals in America had fullfledged hospital-at-home programs. Weeks later it was up to 250 hospitals.

"My phone didn't stop ringing for weeks," he recalled. Even after the pandemic, interest in Hospital at

Home grew for patients with cancer receiving uncomplicated chemo treatments.

"You're also starting to see bone marrow transplants being done at home after that initial three, four, or five days of induction," he says, "because why would you spend another three or four weeks in the hospital with no white cells waiting to get a hospital-borne infection when you can be at home? The rethinking of what happens in hospitals—can we do that at home?—is really starting to happen for a lot of these cases."

THE FUTURE: HOME-BASED ECOSYSTEM

Another proposal on the table deserves attention—the idea proposed by Leff and Dr. Christine Ritchie at Massachusetts General Hospital in Boston. Leff and Ritchie describe a "home-based care ecosystem," a sweeping effort to surround complex older people with complete health care in the home. At its core, such an ecosystem depends upon a radical change that involves, as Ritchie and Leff put it, "shifting away from a facility-focused culture. The US health care system is hardwired for facility-based care. Home-based care is countercultural—a square peg in a round hole of facility-centric care culture. For a decentralized, home-based ecosystem to develop and thrive, leaders in health systems and payer groups must recognize both that home-based care can have great value and that continued focus on 'feeding the beast' of facility-based care is ultimately self-defeating."[34]

Such an ecosystem, Ritchie and Leff suggest, would require "a shift in the facility-focused culture of health care," as well as a new payment system to provide incentives for provider teams to do the work and pay for home-focused supply chains, quality metrics, and shared data.

While Medicare Advantage plans have become sources of significant revenue for private equity firms and the nation's largest insurance companies, Ritchie and Leff see them as having the flexibility needed to "use their health care dollars to meet the needs of patients, including whether and how they pay for home-based care." Accountable Care Organizations such as the Medicare Shared Savings Program can do the same, they observe. "Such plans have systems to assess their populations' needs and can take a long view of patients' health, recognizing that their responsibility for value extends beyond a thirty-day care episode."

The right type of payment is essential to make home-based models of care work; the emergency measures by the federal government during the COVID pandemic demonstrated that such approaches do work and could expand. The speed with which Medicare went from paying providers minimally to treat patients via telehealth to paying amounts equal to in-person visits, demonstrated the potential for rapid government action in the best interests of patients. To see what is often considered a creaky, lumbering giant of a bureaucracy snap to

attention in such a way encouraged (and maybe astonished) the health care universe.

Ritchie and Leff write that "all of the pieces for a
high-value, person-centered, home-based care ecosystem now exist. With integration of payment and regulatory enhancements, targeting of high-need populations,
attention to coordinated and streamlined logistics, technology, and data, and full engagement of social and
behavioral services, the locus of health care will gradually shift to the home."

It is difficult to overestimate the cultural challenge.
The great majority of physicians and nurses in the
United States were trained in a hospital setting. The
shift to thinking about the home as a default setting is
a reach for many in the business. The hospital has been
the physical soul of medicine for about seven decades.
The second foundational element is changing the way
care is paid for. In recent years, a broad consensus has
developed that doctors should be paid to prevent illness,
slow deterioration, and sustain physical and emotional
quality of life for patients for as long as possible (keeping patients "happy, healthy, and out of the hospital," in
the memorable phrase of Griffin Myers, founder of Oak
Street Health). Ritchie and Leff want many players to
join the party, including "payers, health systems, information technology and data analytics organizations,
logistic delivery organizations (such as Cardinal Health
and Amazon), large-chain pharmacies (such as CVS and

Walgreens) . . . and home-based care programs that provide direct care."

De Jonge is supportive of the ecosystem idea, but offers a cautionary note:

> The ecosystem has a lot of spokes in it. But the hub, the core that makes the wheel turn, is a home-based primary care team. Hospital-at-home occasionally comes in, hospice occasionally comes in, as do home health aides, social and legal services. We need the entire ecosystem that Bruce and Christine describe, including matching quality metrics and payment policy. I think the most critical need is to build up the vital primary care hub of that ecosystem.

CAREGIVER BURDEN

We focus on the challenges facing in-home caregivers in greater depth in Chapter 6, but it is important to note here that building an effective home-based ecosystem will require new approaches to supporting family caregivers in the home. Some family members who provide care to an aging parent experience a deep sense of satisfaction. Others, however, are overwhelmed by the experience. Since many families, out of economic necessity, include two-income households, this means the typical caregiver—an adult son or daughter—works a part- or full-time job while simultaneously caring for their own family plus an aging parent.

Eric De Jonge told us that, in his experience, a program where family caregivers are paid to care for a loved one can be effective.

"I have a patient who's eighty years old, quadriplegic from a cervical cord injury, and she has two daughters who serve as her paid aides for sixteen hours a day," he told us. "The DC Medicaid program is popular and progressive and covers paid daily care for disabled elders up to sixteen hours a day. Such programs need to work simply for a family member to serve that role. For families, that goes a long way. Finally, it is essential to provide funding for a full interdisciplinary house call team that has social workers who can coordinate support services, legal, and emotional counseling to family caregivers." De Jonge says the social workers on his teams are "worth their weight in gold."

We cannot help but think that, were he alive today, Dr. Robert Butler would applaud the progress in recognizing and relieving the suffering of older people. It is a proud achievement. Yet, we suspect that Butler would wonder why so many older people continue to suffer? Why, when we as a nation know how to solve at least some of the problem, don't we do more?

They are the 5 percent of Medicare beneficiaries who carry the highest burden of disability and illness. These are our parents, grandparents, great grandparents, uncles, aunts, siblings . . . They are our family members, woven throughout the fabric of American life. They are the other 1 percent. Not the billionaires who constitute the wealthiest 1 percent we hear so much about. They are the 1 percent of the American population and their families who need care at home. This is a message that needs to be shouted from the rooftops, coast-to-coast. The suffering is real, it is common, and it is needless. And that is on us—the rest of us.

CHAPTER THREE

GERIATRIC SYNDROMES

"Absolutely not, I am not accepting this in any way as normal or acceptable."
—Dr. Sharon Inouye

In addition to heart disease, cancer, and countless other diseases and conditions, older people are at risk for and suffer from what are known as "geriatric syndromes," a term that applies to "clinical conditions in older persons that do not fit into discrete disease categories."[35] Common geriatric syndromes include falls, delirium, incontinence, dizziness, pressure ulcers, and problems with eating and sleeping. In this chapter, we focus on the progress in two geriatric syndromes in particular: falls and delirium.

Falls "pose a serious health problem for older persons, occurring in 30 percent of adults over age sixty-five and 40 percent over age eighty . . . Falls lead to functional decline, hospitalization, institutionalization, and increased healthcare costs."[36] The consequences of falls range from minor bumps and bruises to debilitating injuries including hip and other fractures, head trauma, and joint dislocations. The worst falls prove to be fatal.

In hospitalized older patients, delirium is considered a preventable cause of morbidity and mortality.[37] Each year, more than 7 million hospitalized patients suffer from delirium.[38] It is "one of the most common complications for hospitalized older persons, with occurrence rates as high as 50 percent in hospitalized persons. It is consistently associated with increased rates of morbidity, mortality, poorer long-term outcomes, longer hospitalizations, and costlier treatment."[39]

We center this chapter on the work of two physicians and their colleagues—Mary Tinetti at the Yale School of Medicine and Sharon Inouye at Harvard Medical School—who have devoted much of their professional lives to advancing knowledge and identifying preventive measures to mitigate suffering in older people.

"SHE WAS LIVING HER LIFE ON THE FLOOR."

In 1982, while Dr. Robert Butler was launching the Mount Sinai geriatric medicine program, Mary Tinetti was a thirty-one-year-old physician at the University of Rochester Medical Center caring for older adults. Early on, she noticed that many of her patients suffered painful and often-serious consequences from falling. Older people would fall on ice and snow, going up or down stairs, on sidewalks, uneven terrain, in the bathtub, the grocery store—everywhere. Tinetti did not know what to do about this issue and neither, she found, did her colleagues—even physicians with decades of experience. She reviewed the medical literature on falls, but that yielded little insight.

"We were treating their blood pressure and their heart disease, but we had no way to treat their falls," Tinetti told us. She recognized the status quo medical thinking that older adults fall—always have and always will. Nothing much could be done about it.

However, her view began shifting on a frigid winter night in Rochester, New York, in 1982. That evening, Tinetti made her way through the snowy streets for a visit to the home of Alma Davis. Tinetti had not yet met Mrs. Davis, but she was looking forward to sitting down with her to learn about Mrs. Davis's quality of life, in general, and her health conditions, in particular. Tinetti went to the apartment and knocked on the door, but initially there was no response. After several attempts at knocking, the door finally opened to reveal that Mrs. Davis was down on the floor.

"She invited me in and then she just sort of guided herself across the floor and invited me to sit down and have a conversation," Tinetti recalls. "It was shocking. She had fallen weeks earlier and had been unable to get up, so she just basically started living her life on the floor."

Tinetti's best recollection is that Mrs. Davis was in her mid- to late-seventies: "Old enough that it wasn't terribly surprising that she had had a fall and become frail, but young enough that she could still survive for a period of time after that on her own." It was clear that she had been living on the floor for some time—long enough that her hips and knees were completely contracted, rendering her unable to walk.

Tinetti was struck by Mrs. Davis's acceptance of her condition. "She was just so matter-of-fact about it. She was just, 'Well, yeah you know, my clothes are over there, my neighbor brings my food in, and I have it all down here on the floor.' That was probably what struck me the most, how she just was acting like this was the perfectly usual thing that people did. It's amazing how we can, as people, accommodate situations that, from the outside, are like, *How could you ever do this?* I think she was just one of those people who deals with what life sends her. I don't remember her being angry or depressed at all. She ended up coming to the Monroe Community Hospital that had a living situation for people like herself who were relatively independent, but just needed some support with mobility and some activities. She spent the rest of her life scooting around in a wheelchair and got some help with rehabilitation, though she never did walk, but was much more functional and seemed to enjoy her time there."

Tinetti was raised in Flint, Michigan, one of seven children in an Irish-Italian Catholic family. Her parents emphasized the importance of education, so after graduating from high school in 1969, she went to the University of Michigan in Ann Arbor. Central to the ethos of the time was a belief among students that they had an obligation to act on behalf of people who were suffering. Tinetti arrived on the Michigan campus in the fall of 1969 at the height of much of the turmoil over Civil Rights and the Vietnam War. Tinetti joined in some of the campus demonstrations in support of the former and opposition to the latter.

"I think at most campuses in those years—Michigan, particularly, had a very strong political history at that time and I was completely drawn to that. I'd grown up in a strong Catholic family, went to small Catholic schools all my life. To arrive at the University of Michigan at the height of the antiwar movement, it was definitely something that I was very interested in and participated in. And to see young people that were just so devoted. I remember the day when the draft first started and they did the lottery. All the young men on campus found a TV screen and I remember watching with some good friends and people identified where their draft number was, and in a split second, people's lives completely changed.

"When I went to college, I had no idea what I wanted to do with my life. Those times were just so intense, focusing on civil rights and the Vietnam War, that really, I think almost all of our energies on campus were focused on what was happening today rather than tomorrow."

Apart from political activism, Tinetti immersed herself in academics, taking a broad range of courses, including hard sciences. She contemplated the possibility of pursuing a career in physical therapy or law before deciding she wanted to become a physician, remarking that "medicine was a way to sort of combine my love of problem-solving, science, and being able to help."

Over the summer after her graduation, living in the basement of her family home in Flint, she took classes at two different state schools in the area. At night, from 9 p.m. until 2 a.m., she worked in a country- and western-themed nightclub to earn money for expenses and tuition.

"I absolutely loved it," she told us. "It was a whole different group of people that I never ever would've been connected to, and I became friends with people who had very different lives than I had—single mothers raising children trying to make ends meet, different kinds of people. . . . I got to know some of the regular customers and people were very friendly. It really helped my memory because people loved it if you always knew what they drank and you gave it to them without having to ask them again. And I learned some interpersonal things: How do you deal with people once they get a little drunk and they're starting to act up? How do you de-escalate? So, I learned some people skills that stood me in good stead once I got into medicine."

After graduating from the University of Michigan in 1973, Tinetti earned her medical degree there in 1978, and completed her residency at the University of Minnesota Medical Center before arriving for a geriatric fellowship at University of Rochester Medical Center, where she encountered Mrs. Davis on that snowy night. She liked the nature of geriatric challenges. In training at Minnesota, her colleagues tended to prefer acute medicine in the intensive care unit or emergency department, "where there was an acute problem that you can figure out what it is and take care of it." But Tinetti was drawn to the complexity of conditions afflicting older people.

"I liked the slog of more chronic problems where you weren't sure what's going on, you weren't sure exactly what was going to help. You kind of had to think: *Was it disease A that was causing a problem? What role did their home life play?* I really liked the complexity, that's really what drew me into geriatrics."

"CONSEQUENCES OF FALLS RANGED FROM MINOR . . . TO FATAL."

In the winter of 1982, after her first visit with Mrs. Davis, Tinetti sat down to discuss the case with her mentor, Dr. T. Franklin Williams, who succeeded Dr. Robert Butler as head of the National Institute on Aging. Williams said something that altered the direction of Tinetti's life:

"You know I've always wondered about falls and think that there are things that we can do about it. The cause of the fall is treated as something that's not predictable. I think we need more research into why people fall and what we can do about it."

Initially, this struck Tinetti as a somewhat mundane pursuit. But Williams was an impressive man with "a very gentle but persistent way of nudging you, and not taking no for an answer." She was "sort of embarrassed," she told us. "People were studying strange diseases and why study something that was so common? But the more I thought about it, it appealed to me to study something that was right in front of all of our eyes, but we all ignored. Clinically, it was an incredibly common problem that resulted in a lot of consequences for people, functional loss, even death."

In 1982, she organized a study (published in 1986) of older adults living in nursing homes to try and determine what caused some people to fall, and found that there was no single explanation, but rather an array of issues that contributed to falls.[40] She came to the realization that when there is anything off kilter with an older person's vision, hearing, strength, bones, joints, muscles, or mental clarity, the risk of a fall increases.

"Anything that is impaired puts us at risk," she explained. "Any muscle problem, any neurologic problem, any rheumatologic problem, anything going on in your brain, where you take all the information from the environment and decide how to respond safely. *How do I get around that chair? How do I carry this bag up the stairs without falling?* All of that happens in the brain. Anything in the brain can affect it. Also, our cardiovascular system allows us to have the energy to move and stay alert as we're moving around, so any cardiovascular issue can be disruptive."

In her second study, conducted among older adults living in New Haven, Connecticut, where Tinetti had moved, she found that the more cognitive conditions people had, the greater the likelihood they would fall during the next year. The study reported that the risk of falling increased "with the number of risk factors, from 8 percent with no risk factors to 78 percent with four or more . . . We conclude that falls among older persons living in the community are common and that a simple clinical assessment can identify the elderly persons who are at the greatest risk of falling."[41]

Tinetti was somewhat surprised when the *New England Journal of Medicine* accepted her study for publication.

"Most of their articles were very disease specific and this was one of the first papers where they actually looked at a problem in older adults that was not from a disease perspective," but rather was among the issues under the geriatric syndrome heading.

While she learned a good deal from the study, much in the research was unsettling. As she and her team followed up with patients after a fall, they found that "many of them never got back to the level of function they had before the fall, even if it didn't seem like their injuries were that serious. They just stop doing a lot of the things, whether it was driving or going to the store or dancing or volunteering, things that they had done before and really loved to do. And it seemed like they were physically still capable of doing it, but they were afraid." The issue was a loss of self-efficacy, a sense of confidence or belief in one's own ability.

While many falls resulted in relatively minor bumps and bruises, other such mishaps brought terrible consequences. Tinetti found that these serious fall injuries among older adults were roughly as damaging as strokes where many people never fully recovered. While some patients suffer hip fractures or head trauma, for others, a fall is fatal. The message inherent in Tinetti's research to the medical community was loud and clear: older adults are suffering from falls and we are doing very little about it.

COULD SOME FALLS BE PREVENTED?

The next question that Tinetti ruminated on was whether there might be ways that doctors and their teams might intervene to decrease the risk of falling.

"You are not going to be able to do what you need to do in life if you cannot maneuver, get around, and do things," she said. "The idea was not to eliminate falls altogether, but to reduce the likelihood of them." Even a modest reduction would spare many older adults from serious consequences.

Tinetti and colleagues published a second paper in the *New England Journal* investigating "whether the risk of falling could be reduced by modifying known risk factors."[42] Tinetti asked primary care physicians in New Haven if they would allow her team to conduct a study to identify particular risk factors for each of their patients and subsequently intervene where Tinetti's team felt it was necessary. For some the issue was balance or gait, for others it was strength, mental acuity, medication usage, or an unsafe home environment.

"We identified what individual combination of risk factors each person had and then tailored interventions" such as physical therapy, lowering or eliminating medications known to put older adults at risk of instability, or low blood pressure when standing. The results: During the time period of the study 47 percent of those in the control group fell compared with 35 percent of those patients who had exercised, received physical therapy, and reduced their medications. The interventions used in the study decreased falls. The study found that falls "pose a serious health problem for older persons" and "lead to functional decline, hospitalization, institutionalization, and increased healthcare costs."[43]

It was the first peer-reviewed study in a major journal showing that it was possible to reduce the risk of falling, "which was why it was so well received," she said. "I think it helped launch the field." (Since the study was published, more than seven thousand papers by other researchers have cited that particular work.) The results of the study received some notice in the news media, with Reuters reporting that "falls are a major cause of death and disability among the elderly," but that people "can significantly reduce their risk of falling by monitoring blood pressure, taking prescription drugs carefully and using techniques to increase mobility." Reuters quoted Dr. Evan Hadley of the National Institute on Aging, which helped finance the study: "For an older person, this type of fall prevention strategy can mean the difference between being able to live safely and independently at home, or needing nursing home care or other assistance."[44]

While Tinetti and colleagues had demonstrated that adjusting risk factors could reduce falls, she knew from experience that few physicians or other health professionals employed fall prevention strategies. This was not negligence on the part of doctors. Some accepted the commonly held belief that falls were simply not a medical issue. Another problem is that the many risk factors come under the purview of different specialists and other health professionals such as physical therapists. No one clinician is responsible.

Tinetti's next step was nearly a decade in the making. She had demonstrated that with the right interventions falls could be reduced. Now, the question she wanted answered was: Could primary care and specialist physicians, nurses, physical therapists, hospital emergency departments, and home care workers be persuaded to incorporate preventive measures into their standard work? And if so, would it reduce falls? Coincidentally, as she was preparing her study plans, the Connecticut Hospital Association was conducting a survey to better understand the injuries that caused people to visit hospital emergency departments throughout the state. Would it be car accidents? Gunshots? No.

"Far and away, it was older adults having falls that was the major cause of injury in the emergency department," Tinetti said. "They realized what a big problem it was." This surprising finding brought in funding for the study from the hospital association's Donaghue Foundation.

Tinetti and her colleagues selected the south-central portion of Connecticut as the control group, while the Greater Hartford area was the interventional half of the study. In the Hartford region, Tinetti and colleagues advised, taught, and guided doctors, nurses, and physical therapists on fall prevention techniques.[45]

"In our intervention area, we went to primary care practices, some specialists, emergency departments, rehabilitation facilities, home care agencies, any places that took care of people that might be at risk for falling." Identifying fall risk factors and interventions appropriate for each setting—monitoring of blood pressure, physical therapy, and reduced medications—was an arduous, years-long effort.

In addition to Tinetti, the team included Dorothy Baker, PhD, a nurse; Margaret Gottschalk, a physical therapist; and Dr. Mary King, a geriatrician at the University of Connecticut. For three years, they drove around the Greater Hartford area visiting emergency departments, rehabilitation facilities, home care agencies, and clinical practices to train health professionals how to identify persons at risk of falls and how to carry out the interventions that prevent them.[46]

Tinetti was able to show a reduction of 10 percent in fall injuries within the intervention area relative to the control area. The Connecticut study was important news and spread quickly, not only within the medical community, but beyond as well. *The New York Times* reported that "falls among the elderly, a common source of injuries, have largely been considered inevitable. But a [contemporary] large-scale study shows that a combination of adjusting treatment, assessing risk, and educating patients can substantially reduce serious falls."[47]

"THIS THING RIGHT IN FRONT OF OUR FACE."

Twenty-six years had passed since young Mary Tinetti discovered the woman living on her apartment floor. In that time, she had become internationally recognized as a leading expert (if not *the* leading expert) on falls among older adults. Also, during that time, American medicine awakened to the seriousness of falls among the elderly. Awareness of the problem had vastly expanded.

In 2009, in recognition of her body of work, Tinetti was the recipient of a MacArthur Foundation grant. The American Geriatric Society described her as "the first investigator to show that older adults at risk for falling and injury could be identified; that falls were associated with a range of serious adverse outcomes; and that multifaceted risk-reduction strategies were both successful and cost-effective. Her work has transformed the prevailing view of falls as an inevitable consequence of aging to a preventable event with a multidimensional set of risk factors that can be identified and controlled."

Notwithstanding Tinetti's achievements, she experiences her share of frustration. While some subsequent studies have affirmed her findings, others have not. The main obstacle to improved fall prevention, she believes, is the structure of health care in the United States. Physician groups and health systems have been awakened to the risk of falls and often respond by appointing a staff member as "the fall prevention person," says Tinetti.

"But if they are not the people that are making decisions about what medications [older patients are] going to take, or have the ability to make changes to [their] blood pressure, or to make sure that [they] get physical therapy, it's not going to change anything."

As she was analyzing the mass of data from her

Connecticut study from 2004 to 2006, and as she had been caring for older adults for over two decades, Tinetti was struck by a realization similar to what had led her to work on falls. She realized that almost all of her patients suffered from multiple problems, yet the medical status quo at the time was to treat individual diseases and conditions, rather than to consider the whole patient. Like the prevention of falls, it was something that, as Tinetti put it, "We hadn't noticed this thing that was right in front of our face." Her patients did not have just cancer or just diabetes or just heart disease; they had many other diseases as well as geriatric syndromes.

She was swimming against the current of the time when specialists throughout medicine were advocating for the standardization of care, which called for identifying the best practice for any medical condition thus avoiding variation in treatment—an approach that worked well in many cases, especially for those under age sixty-five with one particular ailment. However, things quickly grew complicated when rigidity of treatment was applied to older people.

When Tinetti's Yale colleague Dr. Terri Fried asked older patients with multiple complex chronic conditions to define their health priorities, the majority clearly stated their primary concerns were function and symptom relief.[48]

"When you have multiple chronic conditions," said Tinetti, "what you do for one disease may not help another disease, make another disease worse, and on the whole just treating each individual disease in isolation can cause more harm than good. If you look at older peoples' problems list, there is not one disease..... There are usually three, four, five, or more, and so this concept of multiple chronic conditions or multimorbidity was something that I latched onto. If you give people three medications for their diabetes and four medications for their heart disease and two medications for their depression, you're up to ten medications before you know it."

All of these medications frequently added up to "unclear benefit, potential harm, and a lot of burden to individuals." More medications decreased the likelihood that patients adhered to their designated care plan.[49]

Around 2010, Tinetti added new work focused on the priorities of patients to her portfolio of research on falls. Her new focus was "identifying what decision-making should look like for older adults with multiple chronic conditions. And one of the things that we know is that there is harm in treating all these individual's diseases without thinking about outcomes. We know that, as we get older, we vary in what outcomes we want from our health care. Some people want to live as long as possible, even if it means they're not as functional, or that they have a lot of discomfort, and they're willing to put up with a lot of complicated and bothersome care. And other people prefer to focus on comfort for however long they have. And so, we figured if care was uncertain, potentially harmful, shouldn't we be identifying what was most important to them and align care with what matters to each individual person?"

This has been Tinetti's area of focus, in addition to her study of falls, for about the past decade. The program, known as Patient Priorities Care, helps older people and those with multiple chronic conditions identify the outcomes they want from their health care, and what they are willing and able to do to achieve those outcomes. We explore her work on this issue in greater depth later in the book.

THE DELIRIUM BREAKTHROUGH

Sharon Inouye grew up in a traditional Japanese American family in California with deep reverence for her elders.

"My mother's mother lived with us in our home and my father's parents lived right across the street," she told us. "I grew up pretty much surrounded by older family members my whole life. My aunt and uncle all lived close by on our block and were in and out of our house. The credo in our family was that immense respect went to the older adults in our family . . . My earliest memories were wanting to spend more time with my dad, a very busy general practitioner. He was my hero, my role model. He was the kind of old-time doctor who made house calls and was on-call 24/7 for his patients."

In addition to his practice, Dr. Mitsuo Inouye worked

to get health care coverage for Hiroshima bomb victims, known as the *hibakusha*. "These were American citizens who were living in the Hiroshima area during the atomic bombs," Inouye told us, "or some were people who married American citizens and then came back to the States. Universally, in the United States, they could not get health insurance coverage. My father provided free care to hundreds of *hibakusha,* and he also did political advocacy with Congress trying to get health care insurance coverage for the *hibakusha*."

As a young girl, Sharon spent much time accompanying her father on hospital rounds and, in the process, she developed a love of medicine. "My two brothers had a lot of pressure to become physicians, and I said, 'Oh, but I want to, too!' And my father said, 'Oh no, Sharon, it's not for girls.' My mother didn't work, my grandmother didn't work, and it was kind of the way it was in a more traditional Japanese American family, particularly when your father's a physician. Even though I wasn't encouraged to go into medicine, once my family saw many years later that I wanted to, they were encouraging and supportive."

Her interests shifted over the years, however. She was all about medicine as a child, but in high school, English literature and music ("I fell in love with Baroque music") captured her imagination so completely that they became her obsessions while at Pomona College in Claremont, California. She had studied piano since childhood but in college she became so passionate about playing the harpsichord that she not only practiced six hours per day, she also built her own instrument.

"After about a year of playing the harpsichord I was performing in a small chamber group, and I asked my dad whether I could have a harpsichord," she recalled. "He came to my next concert, saw that I was really serious, and let me buy the $1,500 kit, which was kind of a joke, because I didn't know the first thing about managing a drill or a saw or a hammer, and there wasn't internet and YouTube to teach you in those days. So, I checked out books from the library—I remember the *Reader's Digest* guide to carpentry—and I taught myself how to build a harpsichord in my dorm room. Fortunately, I had a single."

As much as she loved literature and music, however, Inouye also took many hard science courses for pure pleasure. When she was a sophomore, a number of her friends who were juniors were applying to medical school and challenged her to do the same on a dare.

"I said, 'well, number one, I'm not pre-med, and number two, I don't really want to go to medical school.' And they said, 'that's perfect. Let's just see if you can get in.' And so I applied a year early, never thinking I would get in." Within the application essay, she included some poetry she had written "about meditating on life at the top of a mountain."

To her surprise, Inouye was accepted by the University

of California at San Francisco School of Medicine, which meant she had a decision to make. During spring break, she went home and asked to shadow her father at work "to really see and understand what you do." That shadowing experience clinched it—she would become a doctor and leave college a year early.

After graduating from medical school in 1981, she completed her internal medicine residency three years later at the University of California San Francisco and Beth Israel Deaconess Medical Center in Boston. She became a research fellow at Stanford University and then shifted to the VA hospital in West Haven, Connecticut, for training in geriatrics and epidemiology. She would remain at the VA hospital for the next eight years, but in her earliest days there, she was thrown into the deep end. After just a few months as an attending physician for the geriatrics service, she found the issue to which she would dedicate years of her professional life: the suffering inflicted upon older people by what was then referred to as an acute, confusional state, and is now known as delirium.

"SIX PATIENTS ACUTELY CONFUSED ON MY WATCH."

"It was 1985, and I'm now a full-fledged attending physician on the geriatrics service at the West Haven VA hospital," she said. "I'm up to my eyeballs in managing complicated cases, running the geriatric unit, the nursing home care unit, and the intensive rehabilitation unit, which is acute rehab for older adults. I loved it."

She was in charge of care for about forty patients suffering from ailments ranging from congestive heart failure to pneumonia, stroke, and heart attacks. She admitted six older patients to the hospital for treatment of various conditions, including heart failure, lung disease, and cancer. When these six patients were admitted, all were cognitively intact upon admission.

"But within a matter of days, all six of the patients rapidly deteriorated." All six patients became acutely confused.

"Two of them died. Two of them went to the Intensive Care Unit. Two of them went to nursing homes."

These patients seemed to be out of touch with reality, unaware of the date, time or where they were. They were unable to recognize loved ones, sometimes experiencing terrifying hallucinations. One was a man in his seventies admitted with congestive heart failure and treated with what were then the standard drugs, including a mild antihistamine, sleep medications, and anti-ulcer drugs. He was also immobilized because he was on oxygen, and he had a urinary catheter and intravenous lines, so he couldn't get up out of bed easily without a lot of assistance. Basically, he was bed-bound, and wound up getting a urinary infection from the catheter. He developed an acute confusional state.

An oncology patient in his eighties, admitted for chemotherapy and radiation, was treated with nausea and pain medications as well as an antihistamine to protect against the side effects of the chemotherapy drugs. "He became very delirious two days after we started his chemo, and he got an infection and wound up going to the intensive care unit." He died in the hospital.

Inouye was alarmed. She spoke to several colleagues, including more-experienced doctors, asking what was going on with these older patients who were okay when they were admitted but then suffered from acute confusion. She wondered whether she was mismanaging the patients.

"And every single person I talked to said pretty much the same thing—'Sharon, this just happens in older people. Don't worry about it. We see it all the time. It just happens.'"

Inouye found this response almost as shocking as the acute confusional state into which her patients had fallen. "This is *okay* to happen? It's okay for older adults to become confused in the hospital? We don't have to worry about it? And everyone just said, 'No Sharon, don't worry about it. It just happens. We see this all the time.'" Inouye reflected upon these words from her colleagues.

So, you are telling me this acute confusional state is common, in fact happens all the time to older people. That it is "just normal." That the consequences of it are devastating to patients—stays in the ICU, even death are okay, even inevitable.

No, she thought. *Absolutely not, I am not accepting this in any way as normal or acceptable.* She needed to get to the bottom of the issue and started her search in the medical records room. This was in 1985, years before electronic medical records. She pulled the bulging volumes of medical charts for the six patients and graphed out everything that had happened to them since their admission to the hospital. She noted the medications the patients had received, the treatments, procedures, their mental status, immobilization, and anything and everything that had happened to the patient between the time they were admitted until the onset of confusion.

She found that there were common clues to the factors contributing to delirium: for many conditions, patients were prescribed a broad array of drugs, drugs, and more drugs that had side effects of confusion; catheters that caused urinary tract infections; immobilization; complications from medical procedures; many tests or vital sign checks that kept patients from sleeping; new infections; dehydration; deconditioning; falls. Then more drugs were added to treat the new complications and conditions. She plotted the treatments for each patient on graph paper and the shared experience of all the patients emerged as a clear picture. Based on her analysis of the records, she hypothesized that many of the treatments and procedures that were administered to patients contributed to their turn for the worse.

"I brought the graph papers to my chief of staff and said, 'This thing, this acute confusion. It's really serious. *I think we caused this to happen.*'"

She needed to get to the bottom of this phenomenon and toward that end she pursued a full-time research fellowship in 1987.

"I told my professors who were my research mentors that I wanted to study delirium," she told us. "They tried to discourage me. I remember the exact words of my main research mentor, who said, 'Sharon, I don't think I can imagine a more uninteresting area to work on than confusion in older adults. Why would you want to do that? It seems so uninteresting. Can't you find something else to work on?'" But she wore down her mentors and, in the end, they allowed her to work on delirium as her fellowship project.

Inouye went on to become board-certified in geriatric medicine and pursued this as her main field of both clinical and research interest.

"I loved it," she told us. "You cannot find a field that is more challenging than geriatric medicine, where every patient has a minimum of ten chronic conditions and minimum fifteen medications and on top of that, all kinds of social determinants of health, family issues, and living situation issues. You cannot think of a more challenging field where everything is an immense problem-solving maze of some kind. It was a perfect field for a person who loves complexity and problem solving, and a field that really not many people were paying much attention to."

DETECTING / PREVENTING DELIRIUM

From 1987–1989, Inouye joined with physicians at Yale and the University of Chicago to create a simple method for identification of delirium. She organized a panel of experts at Yale to develop the criteria for diagnosis through a consensus-building process. The result was creation of the Confusion Assessment Method. Inouye wanted it to be fast (completed in under five minutes) and easy to apply at the bedside.

"The main thing was it had to be simple and straightforward, and I had to be able to remember the domains in my head at the bedside," she told us. "The diagnostic criteria have to be simple. There are only four things you have to remember . . . Four elements: acute onset, inattention, and either disorganized thinking or altered level of consciousness. You just have to remember those four things and then you rate whether there is delirium—yes or no. It's these four items because they make clinical sense, I can keep them in my head, and that's why I think clinicians will use it."

And they have. The Confusion Assessment Method (CAM) has become a widely used diagnostic tool for delirium around the world, used by physicians and nurses in thousands of hospitals and translated into more than twenty languages. The original article has been cited over six thousand times.

Inouye's next step was to design a protocol to prevent delirium in the first place. "I was working with a very,

very skilled interdisciplinary team, who really taught me about the importance of working with a team—that is, working with nurses, with physical therapists, with nutritionists, with pharmacists, with chaplains, with social workers—all in trying to create a program that would meet the needs of older adults in the hospital."

To maintain cognitive and physical functioning in older hospitalized patients, Inouye's research found that interdisciplinary clinical teams had to target six risk factors for delirium: cognitive impairment, sleep deprivation, immobility, visual impairment, hearing impairment, and dehydration. From 1993–1995, Inouye worked to create the intervention program that came to be known as the Hospital Elder Life Program (HELP). Therapies included games, puzzles, and interaction with staff to help patients maintain cognitive abilities. To promote a good night's sleep, particular attention is paid to preparing patients with massage therapy and providing quiet surroundings and darkened rooms. During their hospitalization, patients are walked three times a day if they are able. Patients receive feeding and hydration assistance as needed. In addition to providing hearing and vision aids, staff members are trained in communication techniques with people who have vision and hearing impairments.

Clinical pharmacists and nurses played a key role in reviewing medications and carrying out interventions to prevent patients from having too many medications and medications that might lead to delirium. Working with the team, Inouye looked back on the original six patients that prompted her interest in delirium in the first place and sought to avoid the mistakes made with those patients. Simple interventions kept patients safe and mentally sharp.

The research found that the HELP program "resulted in significant reductions in the number and duration of episodes of delirium in hospitalized older patients." Overall, the incidence of delirium was reduced by 40 percent. The results of this research were published in a landmark *New England Journal of Medicine* study in 1999.[50] The study was the first to show that a substantial proportion of delirium is preventable. News of the results were published by a variety of medical publications, as well as *The New York Times,* the *Washington Post*, Reuters, even the *Ladies' Home Journal.* The HELP intervention gained widespread use throughout the world as time went on.

More than two decades after it was created, a team of researchers in America and China conducted deep dives into assessing its effectiveness in two systematic reviews and meta-analyses.[51, 52] They concluded that HELP interventions resulted in fewer hospital falls among older adults. In a peer-reviewed paper, the authors reported that delirium "is a common and life-threatening problem for hospitalized older patients. Occurring in 14 to 56 percent of patients, delirium is associated with hospital mortality rates of 22 to 76 percent." The report contained both discouraging and encouraging news. In the former category was the finding that "despite its clinical importance, delirium is unrecognized in 66 to 70 percent of patients and is documented in the medical record of only 3 percent of patients when present." The latter category, however, is the paper's conclusion:

> The Hospital Elder Life Program is effective in reducing incidence of delirium and rate of falls, with a trend toward decreasing length of stay and preventing institutionalization. With ongoing efforts in continuous program improvement, implementation, adaptations, and sustainability, HELP has emerged as a reference standard model for improving the quality and effectiveness of hospital care for older persons worldwide.[53]

The researchers found that HELP was not only effective for prevention of delirium, but also helped prevent functional decline and reduced hospital costs. HELP training, available through the American Geriatrics Society (AGS), provides online intervention protocols, screening and outcome tracking procedures. (The program is now known as AGS CoCare®: HELP).[54]

Ideally, the program is implemented within a provider organization by a team consisting of a physician with geriatrics training; a geriatric nurse specialist;

trained elder life specialists (a special role created for the program, requiring only a bachelor's degree and love of working with older adults to screen and enroll patients, oversee volunteers, and track adherence with interventions); and trained volunteers who work closely to implement the intervention protocols. The program under the American Geriatric Society umbrella adheres to the original interventions Inouye identified decades ago: "Cognitive orientation and stimulation activities, therapeutic activities, sleep enhancement strategies, exercise and mobilization, hearing and vision aids, feeding assistance and preventing dehydration, pastoral care, family support and education, and individualized discharge planning."[55]

Studies indicate that preventing delirium in older patients saves about $1,000 per patient per year for hospital costs and about $10,000 per patient per year in long-term care costs. HELP leads to shorter inpatient hospital stays, a decrease in thirty- and ninety-day readmissions, decreased need for nursing home placement, improved patient and family satisfaction, enhanced nursing job satisfaction and retention, and need for constant companions (sitters).

HELP has been adapted for multiple settings, including acute care, surgical care, emergency department, intensive care, palliative care, home care, inpatient rehabilitation, post-acute care, and long-term care/nursing home settings. HELP represents the original and most effective Age Friendly Health System model of care, and helps hospitals achieve Age-Friendly Health System status.[56]

For all of the HELP program's success, the reality is that there are many more health systems in the United States and beyond that have *not* adopted the program. "Delirium still is often unrecognized, and we don't know how to treat it properly to really shorten its duration, and we don't understand its fundamental mechanisms. Those are all areas where we need to do better," Inouye told interviewer Ira Pastor, founder of Bioquark, a company focused on regenerative biology, during an interview in 2020. Inouye's ongoing work involves exploring how family involvement may improve care for older adults at risk for delirium and seeking to better understand the fundamental mechanisms of delirium.[57]

In 2016, Inouye developed a national and international network to advance research into the field she spearheaded: the NIH-funded Network for Investigation of Delirium: Unifying Scientists (NIDUS). She has shifted her own research to explore more deeply the connection between delirium and dementia.[58]

"This is probably the most important question in the delirium field," she explained to Ira Pastor. "What is the inter-relationship with dementia? Does delirium lead to dementia? We have done studies that have shown that if you do have Alzheimer's disease and you are hospitalized and you get delirious, that your pace of cognitive decline is accelerated by twofold—permanently. Delirium is a risk factor for people developing dementia later, new dementia. And dementia is a risk factor for delirium . . . and so, they are very closely inter-related. Is delirium simply a marker of an at-risk person who will develop dementia, or does the delirium itself lead to actual neuronal damage and acceleration of the dementia? I actually think both things are true." Delirium, Inouye says, is "a marker of a vulnerable brain." Her research indicates that approximately 20 percent of patients—the most cognitively impaired—not only develop the most severe delirium, but also go on to to suffer "permanent cognitive decline."

As principal investigator of the NIH-funded Successful Aging after Elective Surgery (SAGES) study to examine the connection between delirium and dementia, Inouye has led research that has helped to clarify biomarkers for delirium. Through more than 150 studies related to SAGES, Inouye has shown that delirium and dementia are interrelated, and that delirium is a marker for those with a vulnerable brain.[59] Moreover, delirium appears to accelerate cognitive decline in those both with and without dementia.

Inouye believes that "preventing delirium should be low-hanging fruit for those working in dementia. Since we already know how to prevent delirium, applying these strategies to at-risk patients provides an unprecedented opportunity to prevent or ameliorate future cognitive decline and dementia. While the focus has been on developing expensive drugs, we have lost focus on intervening on this important modifiable risk factor (delirium), where we already know what to do and can potentially prevent or delay millions of cases worldwide."

Her call to action: We have no time to lose.

CHAPTER FOUR

AGE-FRIENDLY HEALTH SYSTEMS

WHAT IS AN AGE-FRIENDLY HEALTH SYSTEM?

It all started in 2015 when Terry Fulmer, beginning her role as president of The John A. Hartford Foundation, was asked a question by a board member.

"Terry, what's your big idea?"

Fulmer did not hesitate: "Creating age-friendly health systems," she replied. "Because there's no such thing."

It was a biting reply, but was it true? How could it be that there was *no such thing* as an age-friendly health system in the United States, where aging adults comprise the overwhelming majority of patients? Where the dominant share of payments to doctors, hospitals, and rehabilitation facilities is paid by or on behalf of older people via Medicare, Medicaid, the Department of Veterans Affairs (VA), or commercial insurance companies? In a rational universe, wouldn't a system catering to older people shape its services around that cohort?

Inspired by her mother, Margaret Flynn, who was a nurse and army cadet serving in World War II, Fulmer began her nursing career in the early 1970s.

"I was absolutely in awe of her capacity as a nurse and that had a huge influence on me," she told us. Fulmer preferred caring for older patients in spite of the fact that "they were considered less interesting and less important."

"Ageism was rampant. I focused very much on those frail, older people, and I'm particularly drawn to people with dementia because they're the most vulnerable. They can't speak for themselves, and they sometimes vocalize very loudly, and sometimes can be physically aggressive. I'm drawn to those people because they need me the most."

Once she took the reins at The John A. Hartford Foundation, Fulmer assumed one of the most influential positions in the broad community of professionals focused on improving life for older adults. The Foundation, established in 1929 by John A. Hartford and later joined by his brother George L. Hartford (the family founded the A&P grocery chain), the initial goal was "to do the greatest good for the greatest number." Since 1982, the focus has been on improving the care for older adults. The foundation defines its vision as aspiring toward "a nation where all older adults receive high-value, evidence-based health care, are treated with respect and dignity, and have their goals and preferences honored."

As the foundation's new president in 2015, Fulmer wanted to help reorient the way doctors, nurses, and other providers think about care for older people. Soon thereafter, she laid out her idea to Dr. Don Berwick, the founding CEO of the Institute for Healthcare Improvement (IHI) in Boston, where Fulmer had previously served on the board of directors. Since 1991, IHI has been in the business of applying improvement science to better the quality, safety, and efficiency of health care in many parts of the world. Berwick along with his colleague Dr. Kedar Mate, the current CEO and author of the Preface to this book, were intrigued by Fulmer's challenge to help forge a path toward making health care more responsive and accommodating to the specific needs and goals of older people.

Mate (pronounced "mah-TAY") is a Harvard-trained physician who, while leading IHI, also practices internal medicine at Brigham and Women's Hospital in Boston. Mate is a polymath. As an undergraduate at Brown, he initially majored in American history, changed course and went to Harvard Medical School, worked as a primary care physician in Boston, conducted research, joined the faculty at Weill-Cornell Medical College, and worked alongside the legendary Paul Farmer at Partners-in-Health in Africa and on the HIV/AIDS program at the World Health Organization.

Fulmer approached Mate and his team, asking for their input. She told them that she had "some ideas about what I think the age-friendly health system should be," but she said she needed their guidance "figuring out specifically how this initiative should take shape." And then asked, "Can you go further than that? Can you help me make it happen across the country?"

How could Mate resist? This was a massive challenge, but it also was the sort of work he relished. Begin with the evidence, Mate told Fulmer. What does the existing evidence say about what makes care better for older adults today?

"Let's look for the most-evidenced practice models for care for older adults, and find out how the heck they work," he said. "What are the active ingredients in every one of these models?"

The good news was that physicians, nurses, and researchers across the country had created many successful models of care focused on the needs of older adults. Mate and his IHI colleagues identified about three dozen such programs with good data to support their efficacy. Of the three dozen, Mate's team focused on seventeen programs with strong evidence that they had improved outcomes for older adults.

Among the many examples were the GRACE program (Geriatric Resources for Assessment and Care of Elders) created by Dr. Steven R. Counsell at Indiana University School of Medicine to "optimize the health and functional status of patients living at home while reducing their need for hospital care"[60]; the NICHE program (Nurses Improving Care for Health System Elders)[61] was created by Fulmer in the mid-1980s to have nurses "address common geriatric syndromes and issues" such as cognitive impairment, mobility, transitional care, and polypharmacy; and the HELP initiative (Hospital Elder Life Program), as noted in Chapter 3, created by Dr. Sharon Inouye and colleagues at Yale to prevent delirium and functional decline in hospitalized, older adults.

Mate and colleagues found that the seventeen models of care shared more than ninety common improvement elements. Narrowing down the models, the IHI team found that "there were only thirteen discrete ideas that were separable from each other," Mate told us. These programs had "demonstrated their utility in either a randomized study or in some kind of controlled experimental setup. These were numerous well-evidenced, well-studied, and well-thought-of interventions that have been available for the better part of sometimes two or more decades."

But there was a flaw that seemed to run through many of these programs, the essential problem that IHI has been trying to tackle for a generation and a half, according to Mate.

"It is a problem of systematic application," he said. "These programs are available. They are just not used that much. That is the fundamental failure. When I look at those models, I am not sure that we need more innovative practice models. What we need is better application of the ones that we have." (There were exceptions, of course, including, for example, Sharon Inouye's HELP program which had been adopted throughout the United States and beyond.) The essential challenge, he said, was "model fidelity," a serious obstacle to scaling these models and programs, which had proven effective in their original site, but tended to lose efficacy when tried elsewhere.

"WHAT MATTERS TO YOU?"

Context and history matter. Twenty-seven years before Fulmer came to IHI with her idea, the Picker Institute had coined the term *patient-centered care* to call attention to the need for clinicians to focus on the person as well as the disease. This was an exciting concept and one that was widely discussed in the ensuing years. In fact, it became a mantra of sorts for nearly every physician group and health system in the nation. *Everybody* was patient-centered. But what did it mean in practice? Over time, it seemed to be watered down into a generic marketing slogan.

But then, in 2012, two providers from Massachusetts General Hospital breathed new life into the idea. Dr. Michael J. Barry and Susan Edgman-Levitan, PA, wrote an article in the *New England Journal of Medicine* that changed the vocabulary of American medicine. Barry and Edgman-Levitan noted that standard medical practice often left patients and families "feeling in the dark about how their problems are being managed and how to navigate the overwhelming array of diagnostic and treatment options available to them."[62] They defined two different types of issues requiring medical decision-making. "For some decisions," they wrote, "there is one clearly superior path, and patient preferences play little or no role—a fractured hip needs repair, acute appendicitis necessitates surgery, and bacterial meningitis requires antibiotics."

However, there is quite a different sphere in which decisions are more nuanced, and where "more than one reasonable path forward exists (including the option of doing nothing, when appropriate) . . . Decisions about therapy for early-stage breast cancer or prostate cancer, lipid-lowering medication for the primary prevention of coronary heart disease, and genetic and cancer screening tests are good examples. In such cases, patient involvement in decision-making adds substantial value," particularly "when fateful health care decisions must be made—when an individual patient arrives at a crossroads of medical options, where the diverging paths have different and important consequences with lasting implications. Examples include decisions about major surgery, medications that must be taken for the rest of one's life, and screening and diagnostic tests that can trigger cascades of serious and stressful interventions."[63]

In shared decision-making, patients listen to the

various options from their doctors, weigh the costs and benefits, and then make a well-informed decision centered not necessarily on usual treatment, but on the patient's preferences. Clinicians, Barry and Edgman-Levitan write, "need to relinquish their role as the single, paternalistic authority and train to become more-effective coaches or partners—learning, in other words, how to ask, '*What matters to you?*' as well as '*What is the matter with you?*'"[64]

THE STARS CONVENE IN CAMBRIDGE

The earliest echoes of the age-friendly health system came from the Picker Institute and the journal article by Barry and Edgman-Levitan. But it was supercharged on a steamy Tuesday in Cambridge on August 16, 2016. The region was rich with tourists that week, crowds on the Freedom Trail, aboard Old Ironsides, riding the picturesque Swan Boats in the immersive beauty of the city's Public Garden. Fenway Park was eerily silent as the Red Sox had traveled to Baltimore to beat the Orioles 7–4 (with Mookie Betts hitting two homers and driving in five runs!).

Across the Charles River from Fenway, thirty-one of the most-dynamic innovators in American health care gathered on the fourth floor of a stylish brick office building at 20 University Road on the edge of Harvard Square. The group had gathered to try and answer one of the fundamental questions in American medicine:

When caring for older adults, what is the heart of the matter? How should we, as a society and as caregivers, be thinking about how to better serve the needs of millions of aging adults?

This gathering came forty-one years after Dr. Robert Butler's famous polemical treatise—*Why Survive? Being Old in America*. During those four-plus decades there had been much progress, some of it chronicled in these pages. The expert panel included five leading health systems, principal investigators who had developed geriatric care models, health system leaders, chief physicians and nurses, other clinicians, and federal regulators— some of whom were older adults themselves or caregivers of older adults. They were experts in many different realms.

If some of the best minds in medicine were treated like rock stars, there would have been thousands of screaming fans outside the building that day because this was a *Rolling Stones* moment. Mary Tinetti had come up from New Haven and Sharon Inouye from just down the road at Harvard. Michael Kanter and Nirav Shah had flown in the night before from Kaiser Permanente in California. Bruce Leff from Johns Hopkins, David Reuben from UCLA, Fulmer from The John A. Hartford Foundation in New York, Donna Fick from Penn State, Steve Counsell from Indiana University, David Pryor from Ascension Health in St. Louis, Al Siu from Mount Sinai in New York.

"We showed them the analysis we had done," Mate told us. "We said, 'Here are the thirteen features of what it means to be giving better care for older adults through all of these practice models. We cannot build a spread initiative on thirteen things. Give us the handful that you think are the most important and most valuable elements.'"

The August gathering, along with subsequent discussions, generated what Leslie Pelton, IHI age-friendly expert, called "the most fundamental building blocks to optimal geriatric care—the 4Ms of an Age-Friendly Health System."

> **What Matters:** Know and align care with each older adult's specific health outcome goals and care preferences
>
> **Medication:** [Use] medication that does not interfere with What Matters to the older adult, Mobility, or Mentation
>
> **Mentation (Mind):** Prevent, identify, treat, and manage dementia, depression, and delirium across care settings
>
> **Mobility:** Ensure that older adults move safely every day to maintain function and do What Matters[65]

This framework defines Terry Fulmer's aspirations for aging adults. "I've been thinking about the 4Ms since I graduated from nursing school," she says, referring to them as "the heart of quality clinical practice of older adults." Applying the 4Ms in practice, she says, requires "understanding the goals and preferences of the older person, then assessing their mobility, mentation, medications, and determining any interactions among those four essential elements of healthy aging. For example, can the older person ambulate safely? Is she strong enough to get to the bathroom? Does she have any cognitive deficits? Is there any evidence of delirium or dementia? Are her medications the right ones for her at this stage and are they the correct doses? Understanding these individual elements as well as the complex interplay among the elements is the essence of quality clinical care for older adults."

LESSONS FROM MR. ALFRED MILLER

Let's look at the case of Mr. Alfred Miller. Mr. Miller was a patient within Northwell Health in New York—a warm, cheerful gentleman, and a favorite among the medical staff. Well into his eighties he was managing to live well at home alone with some family support. Mr. Miller faced a number of medical challenges, including high blood pressure, elevated cholesterol, some arthritis, and cardiac issues. Unfortunately, when he was diagnosed with lymphoma, he lost weight, became more frail, took a fall, and was in quite a bit of pain, which led to him being admitted to the hospital. This raised the question: What was the right treatment for Mr. Miller?

Stabilize his blood pressure and cholesterol? Provide chemotherapy infusions on an inpatient basis for his cancer? Treat his pain with appropriate medications?

Mr. Miller was a gift to us, for he taught our teams how to care for older people under the age-friendly framework. We managed his blood pressure fairly easily, but when we treated his pain aggressively, he objected. The medicine made him mentally fuzzy, and he told us that he would rather tolerate the pain than lose what was precious to him—clarity of mind. In conversations, it was revealed that he was uncomfortable with the lymphoma treatments, which made him feel sick and sapped his energy. More importantly, he said, he did not want to be in the hospital for the infusions or any other reason. At home, in spite of the pain, he was able to converse with family members with a sharp mind. He made clear his desire to avoid medications that served to degrade his quality of life. Northwell clinicians stopped both the pain and cancer meds. Here was the key: he was relatively comfortable, functional, and at home—and that was what he most wanted. This is a classic example of treating the patient rather than the diseases, and it gave Mr. Miller the quality of life he wanted.

The desire of aging adults to remain at home is common and heartfelt. Another one of our patients was about to be admitted to the hospital when he objected: He had to be at home with his wife. She needed him. And no matter his condition, and for whatever time remained, he desired above all else to be there with her. To accommodate him, we were able to provide a variety of services in the home—medications, lab tests, infusion therapy, an IV, and oxygen. Telehealth made communications with him and our team simple. There are countless stories like these since the introduction of the age-friendly campaign. At a VA medical center, an older vet was asked what mattered most to him. He paused for a moment.

"No doctor has ever asked me that before. Thank you for asking." Another vet struggling with rehabilitation was asked what mattered to him. His face lit up and he talked about his six-year-old granddaughter. It became clear that thinking about her motivated his rehab efforts. It opened up the patient to more engaged discussions with caregivers.

SHARED LANGUAGE

Age-friendly health care—both respectful and evidence-based—transcends medicine and, at its best, develops the plan of care through the patient's eye. The age-friendly approach encompasses not only the acute care experience, but also the ease of access points at hospitals and physician practices, parking, accessibility, and signage in some of the most-confounding physical environments ever constructed—hospitals.

Fulmer has particular admiration for how the VA health system is working to integrate the age-friendly framework into such a massive system. The VA has gone all in on age-friendly. Kimberly Church, the VA's national leader on the initiative, told us that the power of the 4Ms approach is that it "creates a shared language for how we talk about care for older adults. It simplifies things because care for older adults can be very complex. If we're all working towards aligning care with what matters most and the entire care team knows what matters to the veteran in their own words, it gives the team something to rally behind." The initiative has also simplified and improved communication among staff members and "boosted employee engagement and morale during and following the COVID-19 pandemic. Veterans are really appreciative when we start the conversation with what matters, and then build a care plan that is centered around them," Church said.

In 2023, the VA was in the process of having a what matters conversation with every enrolled veteran. The discussions focused on understanding what brings patients joy, what makes for a good day, what the veteran would like to do more of in life, along with an array of other questions designed to get to know the veteran's core desires and to be able to act upon those desires.

Church's colleague at the VA, Dr. Tom Edes—a key player on the Independence at Home team, whom we wrote about in Chapter 2—has found that the age-friendly framework helps address the shortage of health care workers with geriatric expertise. The simplicity of age-friendly health care clarifies the essentials so that all personnel can readily understand what the overall goal is for veterans. Doctors, nurses, social workers, dieticians, clinical pharmacists, physical therapists, and mental health professionals all readily grasp the essential nature of the 4Ms.

"If we get the physical therapists and the nurses and the doctors in every care setting to follow that framework, that is immensely beneficial in addressing the workforce shortage," Edes told us. "We need people in all disciplines with geriatrics expertise to be champions and to say, 'okay, here's how we do it, here's how we assess mobility, here's how you ask the questions about what matters, here's how you assess cognition.' All those things that will improve that practice. So it's a great workforce extender."

"WHY DON'T WE ASK THE PATIENTS WHAT THEY WANT?"

Our experience adapting the age-friendly system approach at Northwell Health has revealed the inherent power of this deceptively simple idea. We have long been well-acquainted with and connected to IHI as a consultant and advisor to our teams on a variety of different initiatives, including an effective effort to prevent sepsis. In addition, one of us (Northwell President & CEO Michael J. Dowling) previously served as IHI's chair for several years and currently serves as a board member.

In 2014, we engaged an IHI team to help guide our work on improving care for individuals, mostly aging adults, with advanced illness. Our inpatient, outpatient, and skilled-nursing facility teams collaborated with IHI to define advanced illness more precisely: One or more chronic diagnoses, including frailty or dementia plus 2 or more of the following: Declining functional status; malnutrition (unintentional loss of > 10% body weight over the past 3 months); evidence of organ dysfunction; cancer, advanced or metastatic disease.[66] With a working definition, the question much discussed among team members was "what next?" What can we do to ease the burden on these patients? And a key step forward came when we collectively decided to *ask these patients what they want us to do to ease their burden.* Increasingly, the central questions guiding doctors, nurses, and other givers throughout the nation is not *"what is the matter with you,"* but *"what matters to you?"*

Leslie Pelton at IHI explains that age-friendly is not a "highly prescriptive model, but rather a framework that articulates the essential elements of an improvement. It then provides space for local adaption so that it aligns with the local culture." At Northwell, we have identified a metric or an evidence-based tool for each of the 4Ms and we are educating our teams regarding each of the Ms. A dashboard we created within our electronic health records, for example, reports on which patients (including family caregivers) are having "goals-of-care" conversations with their provider and what their outcomes look like. We found at our pilot site (Northwell's Glen Cove Hospital) that having goals-of-care conversations in the emergency department decreased both re-admission within thirty days of being discharged and length of hospital stays—both important improvements.

In the realm of medication, we have a long history of working to manage the effects of polypharmacy, as we described above. Under the age-friendly umbrella, we have revised our electronic health record so that when providers want to order meds, we default to lower dosages at the start. We have also zeroed in on the initiation of three pharmacological repeat offenders: long-acting benzodiazepines (e.g., diazepam), long-acting opioids (e.g., fentanyl transdermal), and tricyclic antidepressants. Additionally, we are working to minimize the use of Benadryl™ (a.k.a. diphenydramine) for non-allergy related reasons, and we have a strong preference with older patients for short-acting opioids recognizing that longer-acting drugs can linger in the system and cause harm. Our newly formed older adult medication safety task force brings together geriatricians, palliative specialists, clinical pharmacists, and nurses to monitor and update changes in best practices for the use of drugs in older patients.

With the third M, mentation, we have focused on preventing delirium by using the delirium diagnosis tools Sharon Inouye developed (as discussed in Chapter 3), the Confusion Assessment Method (CAM) and brief CAM (bCAM).

In the final M, "mobility," Mary Tinetti sounded the global alarm revealing to the health care world the dangers that falls posed to older people (also discussed in Chapter 3). At Northwell, we are using an evidence-based tool (AM-PAC™ "6-Clicks")[67] to assess a person's mobility while in the hospital and screening for fall risk when visiting their primary care clinician. We have found that the Age-Friendly Health System framework breaks through disease centricity and gives providers a more holistic perspective of the patient and their care.

"I think everybody agrees that we needed a more systematic approach to improving care for older people across the entire health system," Fulmer told us. "We now have over 400 peer-reviewed papers in the literature about the work. It's just taken off like a rocket, and that's because it simplifies the approach to care, regularized the care, and promotes quality and reliability. And that means fewer readmissions and often, shorter lengths of stay with better medication plans, consistent delirium screening" and 4M interventions that can be measured.

AN UGLY "ISM"

At Northwell, we are concerned with the issue Kedar Mate raised earlier to explain the need for age-friendly health care. Mate noted that many quality improvement programs around the country have struggled with the "problem of systematic application" or "model fidelity," as he puts it. Is it possible that in the long run the concept of age-friendly care will prove to be immune to that fate?

An important study of the approach at three early-adopter health systems found much to like about the concept, calling it a "compelling conceptual framework for advancing age-friendly care." However, the study also issued a cautionary note that implementation of age-friendly care at these sites was "complex and fragmented." While not altogether unexpected in the early stages of a program of this size and scope, it is nonetheless concerning.[68] That said, among the compelling aspects of age-friendly is its essential objection to ageism in American society. Dr. Timothy Farrell is Professor of Medicine and Associate Chief for Age-Friendly Care at the Division of Geriatrics at the University of Utah. In October 2022, Farrell published an essay entitled "Fighting Ageism with Age-Friendly Care."[69]

Walk into the greeting card section of any department store and you'll find plenty of birthday cards taking jabs at grandma and grandpa for being old, cranky, frail, and always two steps from the grave. Though cards like these are meant to be funny, they still speak to the larger problem of how negatively our society views old age. Ageism—an "-ism" less confronted in the mainstream—has become so ingrained in our culture that it has permeated our health systems, too.

Dr. Robert Butler coined the term "ageism" in 1968, while early on, Drs. Christine Cassel and Diane Meier saw it firsthand in practice. As a society and a profession, we have come a very long way since then. But, as Farrell notes, "older adults are still often excluded from critical research trials" and vestiges of GOMER ("Get out of my emergency room") culture linger. "The 4Ms aim to reduce burdensome and unwanted care by focusing on the health goals and care preferences of older adults . . ." Farrell writes. "These considerate practices may seem obvious, but they are too often left out in favor of a quicker, one-size-fits-all approach." The Age-Friendly Health System framework is changing that and doing so to the benefit of millions of aging adults.[70]

GAME CHANGER

Kedar Mate estimates that the age-friendly framework has been adopted by anywhere from four thousand to five thousand sites of care. There have been more than four hundred publications related to age-friendly work and just about every gathering of physicians focused on geriatric or palliative care includes serious discussion of age-friendly. Earlier in the chapter Mate spoke of the failure of "systematic application" for some very good care models in the United States but age-friendly care has defied the "model fidelity" gravity and spread rapidly. Why has that happened?

"We found a community that largely felt ignored," he told us. "The geriatric community scholars and leaders felt like they were working in the back of the enterprise—the neglected stepchild, not really paid attention to, under-financed. They weren't filling their fellowship spots. I think we found a group that was frustrated enough that they could galvanize around something."

"When we started this, there was no mention of age-friendly health systems. Now, you cannot go to a meeting of the American Geriatric Society (AGS) or the Gerontological Society of America without hearing about age-friendly. When you look at the agendas for the AGS meetings you see age-friendly all over them now, which was not the case before. In some senses, the movement we started we've kind of lost control of, which is a good thing. A lot of what happens is that many other people start to seize upon the effort and work on it without the movement's creators knowing about it. That's what we want. To have that kind of distribution and scale of impact that we're looking for. We need a lot of people in a distributed fashion all together using a similar songbook to do their work from." At the same time, the looseness of the framework allows physicians to tailor the 4Ms to the specific needs of their institutions.

"We're tight on the aims of trying to reach every older adult with better care, tight on the goal of reaching everyone, tight on the four M's, but loose on how those four Ms are put into motion in every practice location," Mate noted. "How you ask or when you ask what matters, just as an example, is largely irrelevant so long as it happens somewhere in the encounter. Whereas if you say you have to ask what matters, use whatever tool you have available, and just catalog which ones have been more or less effective, that kind of flexibility was pretty important to allowing people to make it their own thing without being overly scripted. It wasn't a franchise model where you have to cook the hamburger the same way every time."

Chapter Five

THE RISE OF PALLIATIVE MEDICINE

THE BRITISH INFLUENCE

For much of recorded time, human beings have searched for ways to age well, maintain independence, improve quality of life, and minimize suffering. The quest for the Fountain of Youth and the gift of immortality continues even today.

Not until the middle of the twentieth century, however, did the British begin figuring out how to take care of people in their later phase of life. British physicians saw that life expectancy had increased significantly, creating a new segment of the population living up through the seventies, eighties, and beyond. Seeing this, the British developed modern geriatrics in concert with the creation of the National Health Service (NHS founded in 1948). Lord Amulree (1900–1983), Dr. Marjorie Winsome Warren (1897–1960), and Dame Cicely Saunders (1918–2005) were early aging revolutionists attacking the status quo in different areas of health care to promote aging and the relief of suffering.[71, 72, 73]

Lord Amulree and Dr. Warren incorporated the specialty of Geriatrics and a focus on aging into its country's health system and, in turn, Americans crossed the Atlantic seeking instruction. Today, geriatrics remains the number-one specialty among British physicians—in dramatic contrast to the situation in the United States. Dame Cicely Saunders is the founder of the modern hospice movement and responsible for establishing the culture and discipline of palliative care. She created the first Hospice, St. Christopher's Hospice, and "emphasized the importance of palliative care for patients with terminal illnesses." As we noted in Chapter 1, Christine Cassel, a geriatric pioneer who worked with Dr. Robert Butler at Mt. Sinai in New York, traveled to England for the second year of her geriatric training because there was no such program in America with access to the principles of the fields of geriatrics and hospice and palliative care.[74 75 76]

Fast forward a few decades to the 1990s, when Jack Kevorkian, "Dr. Death," took center stage in the American medical landscape when countless gravely ill Americans were desperate for relief. Many patients wanted to put an end to their lives and sought Kevorkian's help. Kevorkian, a retired pathologist, provided guidance on what was known as "assisted suicide"—lethal drugs administered to people so far outside the normal medical bounds that the deaths occurred in cars, camp-grounds, homes, and in his "rusting 1968 Volkswagen van." He was reviled for his barbarism by the medical establishment, but considered something of a folk hero to some suffering patients.[77] Kevorkian was on the cover of all the major newspapers and magazines, and featured on a segment of *60 Minutes*. His downfall was nearly as rapid as his rise. In the late 1990s, he was convicted of second-degree murder in 130 cases and sent to prison.

A NEW WAY FORWARD

During the 1990s, Rosemary Gibson, a program officer at the Robert Wood Johnson Foundation (RWJF), saw that Kevorkian was a symptom of a deeper problem: the failure of American medicine to provide compassionate care for people with serious illnesses. Not unlike Robert Butler, Christine Cassel, Eric De Jonge, Mary Tinetti, Sharon Inouye, and others, Gibson saw the suffering. More than that, she set out to do something about it. She saw that the people seeking Kevorkian's help felt that they had no alternatives. Their suffering was phys-ical—pain from various cancers and other conditions—but it was emotional as well, a desperate sense of help-lessness. For them, there was no place in the medical care system that met their needs.

While Kevorkian dominated headlines, the RWJF received the findings from a $30 million study it funded designed to identify interventions to provide better care to adults in intensive care units. The *Study to Understand Prognoses and Preferences for Outcomes and Risks of Treatments* (SUPPORT), a randomized control trial, aimed to "improve end-of-life decision-making and reduce the frequency of a mechanically supported, painful, and prolonged process of dying."[78] The two-year study included more than 9,100 adults hospitalized at five major teaching hospitals in the United States who suffered from one or more life-threatening diagnoses including metastatic lung cancer and sepsis with organ system failure. The proposed fix was to enhance "opportunities for more patient-physician communication" as the major method for improving patient outcomes.

Hopes were sky-high, but when all the data was analyzed, the results were a crushing disappointment. The SUPPORT interventions had "failed to improve care or patient outcomes." The findings "shook the medical world," wrote Dr. Kenneth Covinsky, a clinician-researcher from the Division of Geriatrics at the University of California San Francisco.

"The finding that this multi-million effort had absolutely no impact on improving the quality of end-of-life care was stunning."[79]

The combination of the Kevorkian phenomenon and the failure of the SUPPORT study spurred Gibson to find solutions.

"Jack Kevorkian was out there helping people who thought they had no other choice, and this was an enormous embarrassment" to the medical establishment. "These events touched the conscience of good people in medicine, many of whom knew that patients were not well-served," she told us. Gibson believed that the study design did not address the issues that matter to patients. The thesis that "enhancing opportunities for more patient-physician communication" was the remedy for better end-of-life care struck her as simplistic, perhaps naïve. In fact, she was not surprised that the intervention didn't move the needle. The study design did not include attending physicians having conversations with patients and family members.

"Medical school taught physicians how to bring people into the world," Gibson observed. "But not how to help them leave this world."

The work of providing physical and emotional comfort to people suffering at the end of life was an obscure and little-studied subset of medicine. Gibson believed that the RWJF resources, in partnership with selected physicians, could help change this reality.

In 1996, after the SUPPORT study findings were released, Dr. Charles Von Gunten of Northwestern Medicine and Dr. Linda Emanuel of the AMA created an RWJF-funded program called Educating Physicians in End-of-Life.[80] The program teaches palliative care skills in communication, ethical decision-making, and symptom management. Hundreds of physicians turned out for the instructional sessions at medical centers throughout the country, and subsequently demand for training quickly outstripped supply of instructors. As one physician participant said, it was "like feeding manna to hungry people."

Gibson wanted a way to expand the teaching so that physicians would have the skills that would ease suffering for people at the end of their life. She had been following the work of Drs. Cassel and Meier who had established a palliative care consulting service within the department of Geriatrics at Mount Sinai Hospital. Cassel and Meier had been funded initially with modest grants from local foundations, along with financial gifts from a generous patient. Dr. Meier was caring for an osteoporosis patient who "was basically grilling [her] about the work [she] was doing and what [she] was interested in," she recalls.

"I told her about what we were trying to get started in palliative care, and she gave us a large endowment to establish a palliative care institute."

THE PASSION OF DR. MEIER

While Gibson brought philanthropic funding and experience in social change, Meier brought a physician's perspective and a fiery intensity to push palliative medicine forward. Meier had grown up in a mission-driven family where the pursuit of social justice was embedded in her DNA. As a child, Diane spent every summer with her paternal grandfather Frank Meier and they developed a sturdy bond. She recalls Grandfather Frank as a "gentle soul" who rescued many refugees from the clutches of the Nazis in World War II.

"You had to put up, I think, $30,000 and promise that the refugee would not be a dependent on the [US] government and he put up the same $30,000 over and over and over again," she explained. "He rescued many people who otherwise would've been gassed. In the early part of the twentieth century he was a socialist. As he got older, he moved away from that, but he was part of that first- and second-generation of European immigrants who had been burned by autocracies and the tsar and were very focused on equalizing or at least rebalancing power in society. That was the environment in which I grew up."

To Diane's shock, her grandfather died suddenly at home while she was a medical intern. She came to believe that the trend in medicine toward more "organ-centric, disease-centric" care leaves a gap "so that nobody had the full picture of or responsibility for the patient."

She saw something different and more appealing in geriatric medicine which she described as a holistic, integrated field while the rest of medicine was "growing increasingly narrow in the scope of specialties."

"Palliative care was an important, yet under-appreciated part of geriatric medicine," recalls Cassel. "We all eventually are going to die and face that period of time where you don't really want to have aggressive, high-tech, interventional care, and you need a more comprehensive approach to managing symptoms and managing all of the things that make life worthwhile. That's palliative care, and it didn't really exist then."

Gibson asked Cassel and Meier to submit a proposal to RWJF with the goal of establishing an organization that would provide guidance to hospitals and physicians trying to launch and grow palliative care programs. After much discussion, they included a request in their proposal for a grant from RWJF for $200,000.

"Rosemary came back to us and said, 'you need to add some more zeroes'," recalled Cassel. "I almost fell off my chair. That never happens with foundations."

Gibson saw a groundswell among physicians, nurses, and others interested in changing how patients were cared for, and her ambition was to do nothing less than make high-quality palliative care available to every patient who could benefit from it. RWJF was investing in the concept of creating a Center to Advance Palliative Care (CAPC) in addition to Cassel and Meier's ability to carry the program forward.

As Cassel put it, "Rosemary understood better than we did at the time what it would take to actually do that." It would require marbling palliative care throughout academic research, journals, and textbooks, throughout education and training for physicians and nurses, and gaining subspecialty status and establishing quality standards. The challenge was to scale palliative care across American hospitals. Even more broadly, the goal was to begin shifting the culture toward recognizing that living with a serious or life-threatening illness could be managed to ameliorate suffering.[81]

Gibson, Meier, and Cassel collaborated to create a new organization, the Center to Advance Palliative Care (CAPC). The organization's purpose, as Gibson put it, was to "integrate palliative care into the genome of American medicine," an ambition that seemed impossibly large at the start.

CAPC was established within The Mount Sinai School of Medicine and the hospital's Department of Geriatrics, which later evolved to become the Department of Geriatrics and Palliative Medicine. Work then commenced to convince stakeholders at health systems throughout the nation that a new way of tending to older people could make a difference in how patients experienced serious illness. Above all else, the goal of CAPC was to do what Meier and Cassel had established as their mission during their early years of training—to reduce suffering among those of any age with a serious, complex illness.

"One of the first things Rosemary said to me," recalled Meier, "was that I needed a social marketing consultant and a health care financing consultant, and my reaction to both of those recommendations was, 'why?' I'm just trying to help people. I'm not trying to run a business here. I'm not trying to sell something.

That's how naïve I was! Rosemary was able to convey to me that basically what she was asking us to do was run a campaign. We had to identify the key audiences whose behavior needed to change to scale palliative care across American hospitals, and that there were multiple audiences whose behavior needed to change in order for us to do that. Each of those audiences had different motivations, different concerns—payers, C-suites, health systems, policy makers, patients, families, and other clinicians. We did research on each audience to understand what mattered to them, what elements of palliative care would resonate with them to advance palliative care."

When they were launching CAPC, most physicians assumed that patients wanted medical teams to do everything possible to keep them alive no matter what their quality of life was. This was the default setting, but was it in the best interest of patients? Did they know there were alternatives to throwing the medical kitchen sink at the problem? Gibson recalls a physician telling the story of a patient undergoing cancer treatment, which was not working. He understood that the options—clinical trial or hospice—can freeze patients into inaction. He asked the patient, "what is important to you right now?" The patient said she wanted to make sure the animals on her farm were well cared for. She decided not to enter the trial and made the decision to return home where she received home care and eventually hospice care. Many patients have come to fear that the search for a cure can be worse than the disease. The early stages of the palliative initiative anticipated the more-recent age-friendly movement, also asking patients the question— *what matters to you?* rather than the more traditional query—*what is the matter with you?*

Some patients wanted every bit of treatment known to modern medicine. Others sought comfort and time with their families at home. Research showed that the more patients and families reflected on their wishes with physicians, the more they preferred comfort and pain relief. Expensive imaging and other tests had little, if any, value in the final stages of life, but they added significant expense and stole valuable time at home with loved ones.

"What we did with CAPC was to bring palliative care into the acute-hospital setting for a much larger number of people living with serious illness, not just at end-of-life," Gibson told us. The National Institute on Aging defines hospice care this way: "Like palliative care, hospice provides comprehensive comfort care as well as support for the family, but, in hospice, attempts to cure the person's illness are stopped. Hospice is provided for a person with a terminal illness whose doctor believes he or she has six months or less to live if the illness runs its natural course."

"I am not sure if this had been done before anywhere else, especially on the scale that we envisioned and implemented. I use the phrase 'we brought hospice

care in from the cold' as hospitals/academics down-
played its importance. Of course, the Medicare hospice
benefit was designed by family members, nurses, physicians, et al, who could not find a place in mainstream
medicine for palliative care, so hospice was traditionally
community-based. We made palliative care a respected
and essential part of mainstream medicine. *That* was a
big deal."

The CAPC initiative by Gibson, Meier, and Cassel
aimed to enlighten people in the medical community
about the benefits of palliative care. They did so via
"social marketing," says Gibson, "repetition, message
discipline, using the platforms that these different audiences use." They communicated to leaders of provider
organizations and wrote white papers to highlight the
power of interventions by physicians and nurses trained
in palliative medicine, and made in-person presenta-
tions to administrators, physicians, nurses, trustees, and
others who might help advance their cause.

Reaching the general public was a different matter.
When the campaign began, most people had no idea
what palliative care was. Unfortunately, that remains
largely the case even in 2024. As a result, they did not
look for information at palliative care websites. Instead,
patients would go to sites specific to their diseases.

"If you have lung cancer, you Google lung cancer,"
said Meier. "And you end up at the American Lung
Association or at the American Cancer Society and when

you click on those sites, there's tons of information for patients and families." In collaboration with multiple disease-specific organizations, the team was able to get information about palliative care and links to CAPC's public facing website *getpalliativecare.org* included on their websites.

MOYERS ON DYING

Months after CAPC was launched in 1999, it was contacted by producers for journalist Bill Moyers. It turned out that Moyers had gone through the painful experience of watching his elderly mother die, so he set out to do a PBS report on the state of end-of-life care in the United States. This was an important cultural moment.

In earlier times, before the rise of the modern hospital, people often died at home. Over time, the default pattern of hospital care utilized every available technology to forestall death. End-of-life care had become thoroughly hospitalized and medicalized.

Moyers received an RWJF grant to fund a documentary entitled *On Our Own Terms: Moyers on Dying*. The documentary consisted of four, ninety-minute programs in which Moyers interviewed early adopters in the new palliative movement who were demonstrating what quality end-of-life care looked like. PBS defined the series as focusing on the "great divide separating the kind of care Americans say they want at the end-of-life and what our culture currently provides. Surveys show that we want to die at home, free of pain, surrounded by the people we love. But the vast majority of us die in the hospital, alone, and experiencing unnecessary discomfort. Bill Moyers goes from the bedsides of the dying to the front lines of a movement to improve end-of-life care . . ."[82]

As millions of Americans watched, Moyers explained that while "no one wants to think about death . . . there are those tending to the dying who don't want this topic pushed aside. They're working in hospitals, hospices, and homes along with patients to help make sure each person has a 'good death' that fits them, their families, and their culture."[83] *The New York Times* described the documentary as a "panoramic and often profoundly moving vision of how we die in America today."[84]

The response to the film was overwhelming. After the documentary aired, RWJF and CAPC received requests for information from 25 percent of US hospitals. RWJF funding for CAPC was used to assist providers seeking to establish palliative care programs.

"You need East Coast, West Coast, the Midwest—where there are different cultures," said Gibson. "I learned at RWJF that that is what you have to do to diffuse innovation. You bring people together from different parts of the country and different ways of doing things with the same intent and purpose and outcome, and you bring the best minds to it. And that's what we set up as the CAPC network."

In 2002, the long-standing duo of Cassel and Meier was broken up when Cassel left Mount Sinai to become dean of the Oregon Health & Science University School of Medicine. Meier took over as sole director of CAPC and, as Cassel put it, "She really deserves the credit for making that into a phenomenon. CAPC has become a true national force in this country . . . It really has put palliative care on the map."

For palliative care to become part of the genome of American medicine, it needed to be an integral part of medical education and training. Historically, medical schools and residency training programs rarely provided instruction on how to care for people living with serious chronic illness or those in the final stages of life. How could physicians expect to understand and address the complex issues around relief of suffering without training?

RWJF funded an initiative by Dr. Stephen McPhee at UCSF, editor of the textbook *Medical Diagnosis and Treatment,* to recruit academic palliative care physicians in different specialties to write chapters for medical textbooks. Questions about palliative care were added to the US Medical Licensing Exam and RWJF provided support for palliative care physician experts so they could prepare questions for the exam. Gibson observed that "if the questions are on the exam, medical faculty are more likely to teach the content and students and residents more likely to learn it."

By 2006, through the efforts of other palliative medicine pioneers like Charles Von Gunten (an oncologist at Northwestern), the certification boards for medical specialties approved hospice and palliative medicine as a new subspecialty, and nearly a dozen medical specialties—ranging from family medicine and pediatrics to neurosurgery and emergency medicine—signed on to enable their members to pursue additional training in hospice and palliative medicine under their specialty umbrella.

Success of the training "validated that we were creating something needed within the medical profession," Gibson said. "These patients with serious illnesses and their families have a ton of questions, and the palliative care folks would take care of that for the oncologist, for the cardiologist. . . ." All of this progress validated that palliative care was an essential part of patient care.

"With this turning point," said Gibson, "we were on the road to palliative care becoming a permanent part of the health care landscape in the United States."

GROWTH

The growth of hospice and palliative medicine is one of the great success stories in American medicine. From obscurity, this medical specialty has, indeed, become part of the genome of American medicine. After two decades, palliative care is now standard practice in the United States, where programs exist in more than eight out of ten smaller hospitals and over 95 percent of large hospitals. This is particularly significant, notes Meier, because "larger referral hospitals serve the great majority of seriously ill patients . . . It is the larger hospitals that are where the need for palliative care is most acute."

As geriatric medicine was established, and then seeing how palliative medicine grew through the years, we ask: How do geriatric medicine and palliative medicine differ? How do they complement each other? Is palliative medicine a component of geriatric medicine, as Cassel envisioned? Why is the difference between the two specialties sometimes confusing for patients and sometimes for physicians as well?

One difference is that palliative care is not limited to older adults. People of all ages, babies and children included, live with serious illness and need palliative relief from suffering. Many common elements of palliative and geriatric medicine include diagnosis, identification of patient needs and sources of distress, creation of a care plan through shared decision-making, and helping people through their health care journey in a complex and confusing system.

Cassel says that geriatric medicine "includes healthy aging, preventive aging, [and] all of the complexity of care of people from the time they're sixty-five until they are at the end of their life . . . The job of a geriatrician is to keep them going. It's all focused on function. You could have two or three diagnoses—you could have heart failure and diabetes and osteoporosis and osteoarthritis and hypertension, and nobody would ever know it if your medications are managed correctly and if you get the right physical therapy and exercise." A geriatrician's training requires medical expertise in aging, but also incorporates palliative medicine and hospice experience.[85]

The definition of palliative medicine has evolved over the past two decades from a focus on end-of-life care to a broader scope that hews closely to geriatrics' focus on quality of life, functional independence, and aligning care with what the patient says matters the most. In 2010, Meier told an interviewer that palliative care "is about living as well as you can for as long as you can and helping to lift the burden of living with an illness."[86]

"Ask most people what palliative care is, and they'll likely say it's where doctors and nurses help people die peacefully. But they'd be wrong," Meier said later in a 2014 interview. "A persistent barrier is the erroneous belief that palliative care is brink-of-death care. We are working to help doctors and other professionals understand that palliative care is administered simultaneously alongside disease treatment. This remains the biggest hurdle the field needs to overcome."[87]

We are partial to the definition articulated by Dr. Eric De Jonge, who has been caring for older people in the hospital and at home for more than three decades. De Jonge sees geriatrics as "a field that includes doing good palliative care." Since geriatrics and palliative medicine for adults often focus on the same patient population, it might make sense to bring the two together under one roof, according to De Jonge.

"I have actually thought about proposing to my colleagues here at MedStar that we merge our divisions because it's geriatrics and palliative, and it's better to be together."

In 2011, Dr. Helen Fernandez and colleagues developed an integrated geriatrics and palliative medicine fellowship at Mount Sinai, the first integrated program in geriatric and palliative medicine. At least ten institutions have adopted this training model across the nation.

A fascinating article published in the *Journal of Applied Gerontology* in 2021 explored the similarities and differences between the two specialties and, more importantly, sought to fathom why there is confusion between the two.[88]

From the literature reviewed, defining geriatric medicine appears more straightforward than defining palliative care/medicine, which has undergone a shift in how the specialty is defined, perceived, and operates over the last several decades ... While originally focusing on the care of cancer patients, the specialism of palliative care is opening the scope to include other types of patients, which could include older people with comorbidities. This shift in palliative care suggests an increased overlap with geriatric medicine as suggested in a number of the articles reviewed. The boundaries between geriatrics and palliative care are, therefore, unclear. And an overlap has been identified in terms of what geriatricians and palliative care specialists do or could do.[89]

It seems quite remarkable that the study found that "there is *limited understanding between the two disciplines* and what they can offer each other . . ."[90]

A final note about Jack Kevorkian. In 1999, he was convicted of second-degree murder by a Michigan court and served more than eight years in prison. The *New York Times* included this description in its obituary of Kevorkian after his death in 2011: "His stubborn and often-intemperate advocacy of assisted suicide helped spur the growth of hospice care in the United States and made many doctors more sympathetic to those in severe pain and more willing to prescribe medication to relieve it."[91]

The rise of palliative medicine has responded to what many were seeking—*relief*—but much more is needed, particularly better alignment between the two specialties intensely focused on suffering. More cooperation between the two disciplines could help geriatric and palliative medicine specialists, along with the models of care they lead (e.g., primary care, consult services, hospice, home-based care, skilled nursing facilities), to decrease the burden of illness on patients.

CHAPTER SIX

THE INVISIBLE WORKFORCE

How a Small Band of Pioneers Opened America's
Eyes to the Work of Family Caregivers

"OH, MY GOD, HE'S ON HIS OWN."
In the spring of 1973, Susan Reinhard, a newly minted
Registered Nurse (RN) out of the College of New Jersey,
paid a visit to Mr. and Mrs. Smith on a Friday afternoon.
Reinhard was twenty-three years old and filled with
enthusiasm to take the best possible care of her patients.
When she arrived at the residence, she was greeted by
Mr. Smith, a gentleman in his late sixties whose wife
was afflicted with Amyotrophic Lateral Sclerosis (ALS),
commonly known as Lou Gehrig's Disease.

Mrs. Smith had recently been hospitalized and had
a feeding tube, specifically a nasogastric tube—which
extended from her nose into her stomach—inserted.
In the hospital, the task of checking the tube's placement and putting the liquid food into the tube had been

performed by trained nurses, but now, with Mrs. Smith at home, the job was about to become her husband's responsibility. Mr. Smith had never had a moment's medical training in his life. He had been caring for her over a long period and had been able to feed her until she was unable to swallow and led to the placement of the feeding tube. He had been doing the basics—feeding her, bathing her, dressing her, taking her to the toilet, but he had not yet done anything technical.

And so it was on that spring afternoon that Susan Reinhard, RN, would teach Mr. Smith how to perform this delicate procedure of testing the nasogastric tube's placement in the stomach *exactly* correctly—which is typically a nursing task. Reinhard was reminded in the moment of the first time she had checked the placement of the nasogastric tube for a patient while in nursing school.

"I'm thinking, *Oh, my God, I shook the first time I learned how to do this*," she told us. "I had to show him how to listen for air bubbles. . . . You have to start by making sure the tube is in the stomach and not in the lungs. You have to take a bulb syringe and put air in it and then you push the air in and listen over the stomach with a stethoscope to hear the air bubbles to make sure that it's in the stomach. If you do not do this, you will be putting this liquid into the lungs instead of the stomach. And that is really bad, potentially life threatening."

What struck Reinhard as surprising, to say the least,

was that Mr. Smith was expected to learn how to do this procedure in a single home visit. "You're teaching him in front of his wife, and you need to look confident, calm—act like this is a piece of cake . . . I remember how I was trying very hard to be calm myself because I was a pretty new nurse. I was showing him, 'Don't hold the feeding bag up too high because the fluid is going to rush in. Don't force it in. Don't hit the bulb and push it in, just let it go by gravity and keep an eye on her.' And I'm talking to her, too: 'Let your husband know if anything feels uncomfortable because he can just stop it.'"

As it turned out, everything went according to plan on that afternoon, but Reinhard was unsettled. "I felt so strongly that 'oh, my God, he's on his own.' This would be really scary. I could tell he loved her, and I was thinking, *I can't leave him alone over the weekend.* So I gave him my home phone number, which of course you're not supposed to do. I told him, 'if you need *anything*, I will come over the weekend and help you.'" She thought about him all weekend and phoned Monday and to her immense relief all had gone well.

Professional caregivers—doctors, nurses, and others—receive extensive training and education before caring for patients, but family caregivers—daughters, sons, husbands, wives—are often thrown into the deep end of the caregiving pool with little, if any, training.

As she reflected on the experience, Reinhard could not help but wonder about all the other husbands and wives out there caring for a sick and often declining spouse. How many of them had the focus, calm demeanor, and ability to perform a task such as this? She thought about the kinds of things for which family caregivers were given responsibility—tube feedings, injections ("I shook the first time I gave injections"), wound care, sterile dressings, catheters, eye care, ten or twelve different medications (or in some cases even more). Did it make sense to have these various tasks and procedures performed by a well-trained nurse in a hospital, but then leave it up to a barely trained husband or wife to do these things in the home? *How crazy was this?*

Nurse Reinhard made many discoveries in those early years and one, in particular, held true throughout her forty-plus year career to date: she loves caring for aging adults, especially in their homes.

"In a hospital . . . you, as a nurse, are in charge," she told us. "The patient is in bed and the family member is sitting in a chair next to them, they're in a physically subordinate position. In the home, it's completely different. When you walk in the door of their home, you realize that *they* are in charge. You are in their private space, visiting—a consultant to what they have to manage. What's most important is providing as much support and consultation as you can so the caregivers gain the skills and confidence they need." If there are any issues, a visiting-nurse association phone number is provided should a complication or question arise.

Nurses taking on responsibility for patient care in any and all settings—the foundational idea passed down through the ages from Florence Nightingale—was an important part of Reinhard's education. To administer care to patients and learn about settings other than hospitals, the College of New Jersey faculty emphasized that "nurses belong everywhere. That we should not think of ourselves as hospital nurses." Reinhard and other nursing students went to all sorts of different locations—military bases, pediatric clinics, prisons, psychiatric hospitals for the criminally insane.

"That's where my passion for community health came from," she said. "I like hospitals. In fact, I'm the Vice Chair of a health system Board of Directors, but people don't spend that much time in a hospital. They spend most of their time outside of a hospital and their health problems and need for prevention are continuous."

In June of 1973, after graduating from nursing school, Reinhard worked for a time at the Veterans Administration (VA) in East Orange, New Jersey, where she marveled at how patients shared in the caregiving responsibilities for one another. Some of the men, for example, fed those who were unable to feed themselves.

"These were brothers," she recalls. "These were our soldiers who take care of each other."

After the VA, she took a position as a visiting nurse caring for people in their homes. Sometime after she had cared for Mr. and Mrs. Smith, she encountered a patient who was quite poor and suffering from diabetes.

"Her living conditions were not great, and I was very worried that she was going to lose her ability to walk due to the diabetes," Reinhard recalled. "I went to my supervisor and she told me that I could get the doctor to order orthopedic shoes and that Medicaid would pay for it. The patient was delighted. You can imagine she was so happy she could walk around with these shoes." A few months later Reinhard was caring for a patient with diabetes suffering from similar foot problems.

She was thrilled that she had the answer—have the physician write a prescription for the special shoes. But when she raised this with her supervisor, Reinhard had her very first view into the maelstrom that is insurance coverage in the United States. The supervisor told her that no, she could not get the shoes for this latest patient because the woman was not on Medicaid.

The supervisor drew Reinhard's attention to the bottom of the patient charts where an insurance carrier was listed—anything from Medicare to Medicaid, the VA to Blue Cross, and other private insurers. Her supervisor told her that she should always check the coverage to determine what benefits might be available to the patient. At the time, less than a year out of nursing school, Reinhard's world was providing hands-on care for patients. Laws and insurance policies lay in another realm.

"I had no idea what Medicaid was," she told us. "I didn't know what Medi*care* was. That wasn't my job. I thought, *How am I supposed to know about these payment sources?*"

The complexities of payment for health care was absent from her nursing school curriculum. Reinhard had grown up in a medical family—father, a dentist; uncle, a doctor; sister and two aunts, nurses—and while there was much talk within the family about patients and their needs, she never recalled any talk of insurance plans.

"I spent six months in the VA [where] you don't have to think about payment, you're just giving care," she told us. "You don't think about how it's being paid. I think that's true in general whether you're in a hospital or the VA or whatever, you don't generally understand how much things cost. You don't think about who's paying for it. You're just providing your care."

Reinhard realized that she could help one patient at a time as a clinical nurse, but if she could participate in shaping policy, she could have a broader impact. When Reinhard had graduated first in her class from nursing school, the award had been a membership in the American Nurses Association, and she became active in that group as well as in the New Jersey Nurses Association. Starting when she was twenty-four years old, she set out to learn about the world of health policy. Older, more experienced nurses explained how various

insurance plans worked. On top of that, she read everything she could get her hands on. As she learned, she
encountered the often-bizarre nature of payment for
medical services. In her early time as a visiting nurse, for
example, she was required to collect a Medicare deduct-
ible payment of $60.00 from some of her patients.

"Patients were literally handing me cash," she recalls.
This Medicare requirement was changed soon thereafter, which was a great relief.

Within a few years, she was a central player within
the state nursing association to the point where the
group hired her as a lobbyist. In 1983 Reinhard under-
went something of a baptism of fire in the New Jersey
legislature. One of the newer trends was the growing
recognition that some patients preferred to spend the
final phase of life at home with palliative and hospice
services rather than in the hospital. Nurses specializing
in palliative and hospice care, however, felt constrained
by a state law that prohibited anyone but a physician
from officially pronouncing the death of a patient. As a
result, there were increasing numbers of instances when
a family member passed away at home and no physician was available to come to the home for more than
twenty-four hours to make the pronouncement of death.
Under the law, dead bodies could not be moved without
a doctor's pronouncement of death. Thus, families were
forced to live not only with the grief of a deceased loved
one, but also with the terrible anxiety that came with

the deceased individual lying at home in the next room, *rigor mortis* starting to set in.

Reinhard and colleagues wrote a proposal to allow nurses to pronounce death.[92] During the legislative process, the bill was narrowed so that nurses would only be allowed to declare a patient dead in the home, in a hospice, or in a long-term care facility or nursing home. The later irony was that long after the bill had been enacted into law, Reinhard's mother, suffering from advanced congestive heart failure, was at home for the final phase of her life. Reinhard and her siblings took turns staying overnight to care for their mother and Reinhard happened to be with her when she died on Thanksgiving night.

"We were all paralyzed," Reinhard recalled. "You keep wanting to go to the body and you just can't move on" until the body is removed from the home.

Prior to the legislation, the family may well have had to wait throughout Thanksgiving weekend—Friday until Monday—for a doctor to arrive and pronounce death. However, in this case, the hospice nurse arrived and pronounced her dead and the body was moved to a funeral home, which allowed the family members to gather and share stories. Reinhard told her loved ones that on Thanksgiving night, soon before her mother passed away, Reinhard "gave her morphine and tried to roll her on her side because she was uncomfortable. She was coherent and she wanted ice cream and I brought her some vanilla ice cream. And she said, 'thank you,

Jesus, for making ice cream, thank you, Jesus, for making it taste so good.' And those were her last words."

FROM THE BEDSIDE TO THE HALLS OF POWER

Reinhard's profile in New Jersey had risen to the point where, in 1993, she was appointed to the transition committee for newly elected Governor Christine Todd Whitman. Later, she served in the Whitman Administration as Deputy Commissioner of the State Department of Health where she was responsible for all services relating to older adults. She oversaw a staff of six hundred and a budget of more than $2 billion. Governor Whitman also had firsthand experience as a family caregiver for her aging mother, so they both understood the burden on many people, particularly those with low incomes, to work at a job and care for an aging loved one at the same time.

Many family members are reluctant to admit their loved one into a nursing home,[93] but there are limits on what individuals—particularly older people—can do for an ailing person. Sometimes those limits crash head-on into the fervent desire to keep the patient at home and there comes a breaking point where a nursing home is the only viable option.

"Many people say they would rather die than go to a nursing home," says Reinhard. "During COVID, people did go to a nursing home, and many died. I'm not against nursing homes, but I am against having no choice. In

New Jersey, what we promoted was choice. *Long term care, you decide where.* That was the slogan. We were letting people know they have this choice, some options for the one that they're caring for." And options, including respite services and adult medical daycare, Reinhard believes can be the difference between keeping loved ones at home versus turning to a nursing home.

Governor Whitman recognized that in a quarter of the state's households someone was caring for an aging relative, and nearly two-thirds of these caregivers were also working full- or part-time. In many of these cases, caregivers were doing everything possible to keep their loved one out of a nursing home. Whitman said she wanted "a New Jersey where no one has to go into a nursing home unless they absolutely have to be there and want to be there." Whitman and Reinhard created and expanded services so people wishing to remain at home could do so. The Jersey Assistance for Community Caregiving (JACC) program sought to "divert or delay" nursing home admission by strengthening participants' networks of informal caregivers and to maximize autonomy by providing participants with the opportunity to direct their own care, including hiring their own providers, if desired. In each case, a care manager would collaborate with an older person to serve the person's particular needs.

"With the participants' needs, desires, and goals in mind, the plan of care specifies additional services to be delivered," which could include chore services, home modifications, meal service, respite care, social adult day care, and transportation services.

Reinhard had been prescient. During the 1980s, while working toward her doctorate, she had written a paper making the case that family caregivers in some situations should be paid for their work by state government. She and Whitman instituted just such a program in New Jersey.

In 2023, AARP reported that 21 percent of Americans—about 53 million adults—served as caregivers to family members or other loved ones. Sixty percent of these people also work full- or part-time. One in four find it difficult to take care of their own health (23 percent) and a similar proportion report caregiving has made their own health worse (an additional 23 percent).[94] The strain on many caregivers leaves them feeling as though they must be on call 24/7, which can leave them feeling "engulfed" by never-ending tasks. While caregivers typically do the very best they can to help their loved one, the reality is that most caregivers never receive any training at all over the entire course of caregiving. This is concerning and, in some cases, dangerous.

Nevertheless, millions of family caregivers derive enormous satisfaction from the role. The American Psychological Association (APA) also takes pains to cite a 1999 study[95,96] which "found that 44 percent of the spouse caregivers in their sample reported 'no strain' in association with caregiving tasks" and a 2014 survey reported that "83 percent of caregivers viewed it as being a positive experience."[97]

Reinhard's crusade to gain recognition for the needs of family caregivers gained some traction during the 1980s and into the 1990s. She found like-minded allies, including Terry Fulmer, President of the John A. Hartford Foundation, which has funded multiple programs to improve the ability of family caregivers to provide the support needed by their loved ones. Others were pulling in the same direction as Reinhard and Fulmer, including The Family Caregiver Alliance and the National Alliance for Caregiving. Most progress in policy and programs has occurred at the state and local levels, but in 2000 Congress stepped up and enacted the National Family Caregiver Support Program and in 2018 passed the RAISE Family Caregivers Act to create a national strategy on family caregiving. In the 2000s, the pace of change accelerated in part due to demographic shifts.

"For one thing, more people are caregivers now because people are aging, and people with disabilities are living longer," Reinhard said. "As Rosalynn Carter famously stated: 'There are only four kinds of people in the world—those who have been caregivers, those who are caregivers, those who will be caregivers, and those who will need caregivers.'"

AARP COLOSSUS

Reinhard's breakthrough opportunity to shift public perception and public policies toward family caregivers came in 2007, when she joined the colossus that is AARP as director of the organization's Public Policy Institute. Prior to her arrival, AARP staff members were working on a project called *Valuing the Invaluable,* first released in 2007 and updated every few years, most recently in 2023. This effort attempted to spotlight the work of family caregivers by assessing its economic value. The 2007 *Valuing the Invaluable* report asserted that "unpaid caregivers' contributions are not only the foundation of the nation's long-term care system but an important component of the US economy, with an estimated economic value of about $350 billion in 2006. The figure of $350 billion is based on an estimated thirty-four million caregivers age eighteen or older who provide an average of twenty-one hours of care per week to adults with limitations in daily activities."[98]

By 2021, the "estimated economic value of family caregivers' unpaid contributions was approximately $600 billion, based on about thirty-eight million caregivers providing an average of eighteen hours of care per week for a total of thirty-six billion hours of care, at an average value of $16.59 per hour. This conservative estimate does not consider the financial cost of care (out-of-pocket and lost wages) or account for the complexity of care provided (i.e., medical/nursing tasks).

This $600 billion estimate for 2021 is up from $470 billion in 2017 and continues a twenty-five-year trend of increasing economic value."

AARP's campaign to raise awareness of family caregivers was boosted by its analysis, widely covered in the news media regarding the "invisible workforce." The 2023 report included first-person thoughts from several caregivers in the United States describing their lives.[99]

Carri, early thirties, Texas, caregiver for her mother and grandfather: "I am a full-time caregiver for my grandfather and part-time for my mother. Grandpa has Parkinson's disease and prostate cancer. Mom has multiple myeloma . . . I left my entire life, a life I loved, in Nashville, Tennessee, to be a caretaker in Texas. I am so grateful to have this time with my family, and I truly believe that caretaking is a calling . . ."

Angie, early sixties, Pennsylvania, caregiver to her parents and then her in-laws: "I cared for my dad for two years in his home . . . I cared for my mom after Dad died, in my home . . . I now care for my in-laws in their home. I quit my job to relocate across country. My husband closed his small business."

Roger, mid-seventies, New Hampshire, caregiver to his father for seventeen years: "Me and my wife worked, so we had to take turns to come and check in on him . . . The last five to ten years of my father living with us, my dad had multiple accidents. In 2010, my father had a heart attack and his needs increased . . . He could not bathe himself, wash clothes, or feed himself. He had multiple daily living needs. Everywhere we went, we wanted to bring him along—and so we did."

Ayda, early sixties, Florida, caregiver to her father: "My father is bedbound . . . On the weekend, I cannot go anywhere. I have to miss family events because of this. [It] has taken a great toll on my health. Sometimes I just break down and cry because I have to pay out of pocket to get a private caregiver for a few hours to go to church, [and] grocery shopping. . . . I don't have a life for the past four years when my stepmom passed."

SPOTLIGHT ON THE "INVISIBLE WORKFORCE"

When Reinhard started her career in 1973, she called attention to the need to provide support for family caregivers, which was rare at the time. That is not to say that Reinhard was alone in her crusade. There were other nurses, as well as social workers, and researchers who had dug into the topic enough to understand the essential role family caregivers were playing in tens of millions of homes across the United States. Largely, however, challenges for caregivers was seen as a private family issue, rather than a matter for health and public policy.

Yet, by 2015, awareness of the role of family caregivers had increased by an order of magnitude. Everybody, it seemed, was talking about it. By the time AARP released its report *Valuing the Invaluable* in July of 2015,[100] the health care and public policy worlds were paying close attention. The press played a significant role in highlighting the needs of family caregivers, particularly as an increasing number of journalists were taking on responsibility for an aging parent and writing about the experience.

Since 2000, the publishing marketplace has been flooded with books targeting caregivers and their challenges. The 2014 book by surgeon and the *New Yorker* staff writer Atul Gawande, *Being Mortal: Medicine and What Matters in the End,* became a bestseller and hot topic of conversation within the health care community and beyond. A consensus had developed that caregivers in the home were providing care of significant medical and economic value and that many of these 40 million men and women needed help—in many cases, a lot of help.[101]

The "invisible workforce" as it had been called, was now in the spotlight. Susan Reinhard and others had succeeded in calling the nation's attention to the challenges facing family caregivers in the home in much the same way that Mary Tinetti alerted the world to the risk of falls and Sharon Inouye sounded the delirium alarm. Reinhard's journey is akin as well to the Academy

Team that succeeded in advancing the concept of home-based care from virtual oblivion into the mainstream of American conversations in all sectors.

HOME ALONE

Reinhard's long-held assumption was that the clinical care burden on family caregivers was a major issue—that millions of untrained people caring for patients with complex needs were performing tasks that nurses and other health professionals typically performed only after extensive training. And these caregivers were largely left on their own to figure out how to provide care.

There is ample anecdotal evidence on the topic, but limited research. The American Psychological Association has noted that most research has "emphasized caregiving burden and the potential negative effects of caregiving stress on mental and physical health."[102] Steven Zarit, PhD, reports that "research has shown that intense caregiving situations can be harmful to a caregiver's health and well-being. Caregivers in these situations have higher rates of depressive symptoms, anger, lower positive emotions, greater health problems, and higher mortality than age- and gender-matched individuals."[103]

Zarit notes that social and economic changes in recent generations have altered the world of family caregiving. In the past, he observes, "family care has often been taken on by women in the family, specifically daughters and daughters-in-law. Over the last fifty years, more women entering the workforce has raised the challenge of how to maintain employment while providing care to an elder while, in some cases, also taking care of children. Smaller family size also means there are fewer offspring for providing care to a parent. One benefit of increased longevity is that survival of both spouses in a marriage means that if one of them becomes disabled, the other will usually be able to provide care. There are probably more spouses than daughters now providing care, although their own age and health sometimes is a limiting factor."

Reinhard found Zarit's work informative, but she wanted to go deeper. In 2011, she proposed conducting a study to explore in depth what was actually going on in these homes—what family caregivers were doing on a day-to-day basis. Reinhard teamed up with Carol Levine, director of the Families and Health Care Project of the United Hospital Fund, who had experience as an in-home caregiver for her husband. While Reinhard proposed the study, Levine named it *Home Alone*. Reinhard had funds left over from a John A. Hartford Foundation grant and was given permission by the foundation to use the money for the *Home Alone* study. *Home Alone* was published in October of 2012 and gained the widespread attention Reinhard and Levine had hoped for.

The study showed that 46 percent of the nation's 42 million caregivers were not only performing basic tasks, activities of daily living such as bathing, dressing, shopping, or cooking, but also performing complex medical tasks for which, in most cases, they had received little if any training.

"This included giving multiple medications, injections, eye drops, ear drops, wound care, catheter care, special diets, and the list goes on and on," Reinhard told us. "The study documented this huge unmet need for family caregivers in terms of learning how to do this complex care.... At least half of them said they were worried about making a mistake and they felt they had no choice but to do this.... The message was, 'wait a minute, family caregivers are doing more than you think they're doing.' And more important, they need help. No one is teaching them how to do this and they're scared.... They just feel they have to do it but they're scared. Family caregivers are doing a lot that I always said that made me as a nursing student tremble. When I had to give an injection—are you kidding me? And *we just expect them to get able to do this?*"

As a result of the study, AARP drafted the CARE Act (Caregiver Advise, Record, and Enable) for states to require hospitals to ask patients of any age and diagnosis if they have someone who will be helping them with their care once discharged. If yes, does the patient want to include the caregiver(s) in the health record? If yes, the caregiver must be offered instruction in any expected care they will be providing.

Tangible signs of progress were everywhere. In 2013, Medicare began paying for care when aging adults were transitioned to or from a hospital. In 2014, the Institute of Medicine established a Committee on Family Caregiving of Older Adults. In early 2015, a Congressional Caucus was formed "to bring greater attention to family caregiving and help people live independently, educate Congress on these issues, and engage them on a bipartisan basis to help develop policy solutions." AARP reported that a "federally funded Research and Training Center on Family Support is bringing together experts in aging and disabilities to advance a coordinated approach for research, policy, and practice to bolster family caregivers." Some states required hospitals to provide instruction to family caregivers on medical tasks such as injections and wound care.[104]

There have been many initiatives in quite a number of states which have served to recognize the role of caregivers and provide services that help ease their burden of work. Eleven states and the District of Columbia have enacted paid family leave programs, and more than forty states have laws supporting family caregivers during the crucial transitions when their loved one is being moved into or home from a hospital.[105] Many of the major advances have been at the federal level. The Veterans Administration expanded a program of services for family caregivers.[106] The federal CARE Act has spread to over 40 states to improve caregiver identification and involvement in a loved one's care plan creation.[107]

In 2019, Medicare Advantage plans included more services for family caregivers and two years later, the American Rescue Plan Act of 2021 broadened the availability of home-based services.[108] Only a few decades ago, there was little awareness of the reality facing family caregivers. But by 2024, it was near the top of every health care to-do list throughout the nation.

The broader question going forward is how to do better at supporting these caregivers—including training for increasingly complex care in the home? As awareness of the challenge has grown in America, so, too, has the burden, largely due to greater complexity of care family members are asked to provide. As awareness grows, so does pressure on family members. People are living longer and using hospitals for much shorter periods. Not so long ago, many older people would remain in the hospital for some time, then be transferred to a skilled nursing facility for rehab and only then go home. Now patients go from hospital to home, often with the need for sophisticated care. Family caregivers are no longer just helpers, but essential players in determining patient outcomes. This is nothing less than a radical and somewhat-frightening development.

Too often, caregivers are ignored, left out of the process of establishing a care plan. Yet, the literature shows if we involve the caregiver early on in the discussion, the care plan will be more likely to produce a positive outcome for the patient. In the medical community, there is growing recognition that the family caregiver needs to be included and consulted as part of the health care team, and trained as such. This is, by any definition, a significant advance over the status quo of just a couple of decades ago.

CHAPTER SEVEN

VALUE-BASED WILDFIRE

A PIPE DREAM?

When exploring the evolution of American health care from thirty thousand feet, we are reminded to pause and look back through these pages at the patients whose needs define the mission. In Chapters 7 and 8 we explore the major government and corporate initiatives intended to help literally tens of millions of aging adults including the Patient Protection and Affordable Care Act, Medicare Accountable Care Organizations (ACOs) and Medicare Advantage (MA).

As we examine the scope of these efforts, picture Dr. Eric De Jonge's visits with steel worker Paul Stankowski. What would a delivery system look like that would effectively serve Mr. Stankowski's needs? Let's also recall that snowy night in Rochester, when thirty-one-year-old Dr. Mary Tinetti arrived at an apartment to find that Alma Davis had fallen and been living on the floor for weeks.

What would a delivery system that served Ms. Davis's needs look like? Or for those six older patients whom Dr. Inouye admitted to her department who then rapidly, and inexplicably at the time, degraded?

Dr. De Jonge and his colleagues succeeded in getting the Independence at Home program integrated into the Affordable Care Act, Dr. Tinetti shone a global spotlight on the perils of falls among older people, and Dr. Inouye created methods to diagnose and treat delirium. The work by these doctors directly responded to the needs of their patients. More than that, the achievements of these physicians have benefited millions of people throughout the nation and around the world. The same holds true for the palliative care campaign by Dr. Diane Meier and Rosemary Gibson, and for nurse Susan Reinhard's work training a caregiver how to check his wife's feeding tube placement. All of this work, to one degree or another, falls under the age-friendly initiative conceptualized by the John A. Hartford Foundation and the Institute for Healthcare Improvement.

We need to keep these patients and physicians in mind in the next two chapters as we ask some basic questions: Do ACOs and Medicare Advantage plans respond to the needs of these patients? Do they improve care? Reduce suffering?

The men and women we write about in this book are like-minded descendants of Dr. Butler, the first director of the National Institute on Aging. Although Butler died on July 4, 2010, at age eighty-three, the power of his movement has not faltered. In fact, just about one hundred days prior to Butler's passing, President Obama signed the Affordable Care Act (ACA), which accelerated the momentum of Butler's movement.

In this chapter, we focus on the programs spawned by the ACA, particularly Accountable Care Organizations. The government reported that by 2023 there were six hundred ACOs—teams of physicians, nurses, and others—caring for 13 million patients. We convey the story through the work and viewpoints of Drs. Don Berwick and Patrick Conway, both of whom played important roles at Center for Medicare and Medicaid Services (CMS) creating experimental ACOs.

In this chapter, we also introduce Dr. Rick Gilfillan, who played a pivotal role in the ACO world. Gilfillan had an interesting upbringing in a Roman Catholic family in Lincoln, Rhode Island. He noticed everything: how persistently his mother—who graduated from college the same week Rick graduated from high school—pursued education; how his father's work, as leader of the Urban Center in Providence, was all about helping people from poor backgrounds get into college. He also took notice of the caring spirit of Father Sheldon, the parish priest, who coached him in basketball and baseball.

Later, Father Sheldon left the priesthood and started a new career as a community organizer in the tradition of Saul Alinsky, the radical thinker who worked to organize the poor to fight for better housing, jobs, and city services in Chicago. As a student at Georgetown in the late 1960s, Rick worked with Sheldon for a summer implementing Alinsky's principles for Navy Yard workers. "I learned about principles of community organizing," Gilfillan told us. "The power of protest, building coalitions, supporting leadership and action by people in the community."

Gilfillan went on to the Georgetown School of Foreign Service where he immersed himself in the study of history, economics, political science, and philosophy. He then earned an MD at Georgetown and an MBA at Wharton. Why a business degree on top of an MD? "I had no understanding of how business works and how the best businesspeople think and operate . . . I wanted to be a more effective change agent with general management to break out of the clinical-only track." He served as CEO of both Geisinger Health Systems HMO and insurance companies, and as CEO of Trinity Health, which operated across twenty-two states.[109]

Like Gilfillan, Dr. Don Berwick was in search of the holy grail: better health at lower cost. His career was pathbreaking at virtually every step. He worked at Harvard Community Health Plan (HCHP), founded in 1969, an innovative Health Maintenance Organization, and exactly the kind of community medicine Berwick wanted to practice. The group was founded by Robert H. Ebert, dean of the Harvard Medical School, and led by Thomas O. Pyle, a management consultant. HCHP "was created to revolutionize the way in which medical care would be financed and delivered in Boston. In partnership with Harvard Medical School, this prepaid group practice with its salaried physicians would serve the sick while also providing preventive care to healthy members of the practice, with its patients coming from all sectors of society."[110] This description from HCHP foreshadowed the major changes that would come four decades later with President Obama's Affordable Care Act.

Better quality care at a lower cost? Was this a pipe dream? In fact, research at the time showed that the nation had a problem with *overuse, underuse,* and *misuse* of medical care: *Overuse* of tests and procedures that patients often did not need—driven by fee-for-service billing, where doctors were paid for the volume of care rather than the value of care; *underuse* of the kinds of preventive interventions that would keep people from getting sick in the first place; and *misuse*—preventable medical errors that added billions to the cost of care in the country.

HCHP was structured to reduce unnecessary tests, procedures, and errors while increasing preventive measures. Patients prepaid for their membership, which gave doctors freedom to practice without worrying about running on a fee-for-service treadmill to make ends meet. Prepayment was *freedom*—freedom from the miles of administrative entanglements that drive doctors crazy and steal time away from patients; freedom to focus on preventive measures that would keep patients healthier and out of the hospital. HCHP had joined the managed care revolution along with the likes of Kaiser Permanente, Health Partners in Minnesota, Group Health Collaborative in Seattle, and others. A key feature of these systems is that they employ salaried physicians, freeing doctors from volume-based fee-for-service pressure.

Dr. Sachin Jain, President and CEO of SCAN Group & Health Plan, makes the important point that "prepayment affords clinicians and delivery systems the freedom to fund whatever is necessary to truly manage a patient's medical condition . . . a 'whatever it takes approach' to achieving outstanding outcomes for patients—thinking beyond traditional healthcare delivery models to deliver common-sense solutions to patients. Whether purchasing transportation to appointments for patients who otherwise might miss them or purchasing air conditioners to keep COPD patients out of the hospital, true value-based care means thinking beyond the traditional boundaries of healthcare payment and delivery."[111]

The founders of HCHP sought to do what Jain described. They worked in tight-knit primary care teams and coordinated closely with specialists. The whole idea, Berwick told us, was to "keep patients well enough to avoid hospitalization [and] out of the clutches of the sort of technocratic system." Tom Pyle, the HCHP CEO, recalls Berwick was a driven taskmaster who "pushed the hospitals and doctors affiliated with [the Plan] to improve care for patients . . . by setting performance benchmarks and by standardizing treatment based on best-practices research" and measuring the quality of patient outcomes—whether patients received regular mammogram exams, statins to prevent heart disease, and colonoscopies to detect cancer, for example.

This approach was relatively new to health care, but Berwick, who was appointed vice president of HCHP quality-of-care measurement in 1982, took to it with enthusiasm, exploring the ideas of improvement engineers in industries outside health care. He immersed himself in the work of W. Edwards Deming, the quality management guru, and recognized that measurement could be helpful, but was by no means sufficient to improve health care quality. Berwick measured and then measured some more.

"We measured infant mortality, infection rates, waiting times . . . ," Berwick told us, "failure to follow-up tests, endless numbers of measurements and endless conflict because the process itself was not welcome in the organization by the workforce. After all, [the medical teams were] working as hard as they can and I'm showing up with reports saying that people are waiting too long, or that infections are up, or whatever." But the old way of thinking was still in place: that variation in performance reflected individual effort, rather than seeing performance as a property of systems, which can and should be improved.[112]

This was a noble attempt, but with fraught results. To begin with, physicians at the time were unaccustomed to being measured. They were working hard, and the idea of accountability was difficult for them. Instead of using the systems thinking that was suggested in Dr. Deming's work, they took measurement personally. Making matters worse, many of Berwick's measurements portrayed a gloomy picture of how the health care system was often performing. In a room full of doctors, passing out quality data showing variation in patient satisfaction among physicians was humbling, perhaps even infuriating. One doctor balled up the report and threw it at Berwick.

Yet the ability of Health Maintenance Organizations (HMO) to control costs made them popular with insurance companies and employers eager to get medical spending under control. The popularity of HMOs like HCHP grew rapidly through the 1970s and '80s to the point where, "between 1990 and 1995, the number of public and privately insured enrolled in [an HMO] grew from 36.5 million to 58.2 million, and by 1995, the majority of Americans with employer-based health insurance were enrolled in some form of managed care plan."[113] Most encouraging for Berwick was that, by the middle of the 1980s, the HCHP model was working reasonably well. Results included better clinical integration,

15 percent lower costs, and high patient satisfaction. It was a defining moment in American medical history—until it blew up in everyone's face.

MAYHEM

"All hell broke loose—everything went sour," Berwick recalls. "Large insurance companies entered the prepayment arena, saying they could do what HCHP and others were doing. The insurers said 'we can do that, we'll take risk, we'll take pots of money, and we'll manage populations.' But they had none of the culture or the commitment or the nonprofit ethos of the great HMOs—and they took over. That's when the public became enraged about restrictions on care. These were financially driven entities run by people who've never seen a patient, businesspeople responsible to stockholders . . ."

There were certainly cases where denials for care were appropriate; where the denial not only saved money, but also made sense clinically. A well-documented problem of overuse of medical services, driven in large measure by fee-for-service, was prevalent. Nonetheless, the restrictions on care by insurers claiming to "manage care," in the eyes of doctors and patients, crossed a line.

By the early 1990s, for a growing number of patients, "managed care" was a pejorative term describing the experience of being boxed in by narrow networks and muscle-bound gatekeepers who prevented patients from accessing care they wanted. In many cases, workers insured by their employer's health plan had one choice for coverage, leaving millions of workers feeling trapped in a system where the doctor they preferred was unavailable and where seeking out-of-network treatment involved bureaucratic appeals that were frequently rejected.

But the newly empowered insurance companies could only go so far. Soon a howling chorus of complaints from employers, workers, individuals paying for their own care, and government officials made clear that the insurance companies had crossed a line. While there was little disagreement about the need at the time to curb rising costs, the ham-handed way in which it was done made millions unhappy. The response to the denials of coverage, the gatekeepers, the inability to go outside the network for the right specialist, and all the rest was predictable: it bound together patients and doctors to fight back, shaming insurers for their overreach.

"NOT-FOR-PROFIT ETHOS"

Why had the insurers been so tone-deaf? How was it that they believed they could get away with alienating both patients and doctors? Part of the problem was corporate arrogance, but another factor, in Berwick's opinion, was a lack of *not-for-profit ethos. Maybe this is romantic, but the role of the clinicians, the physicians, and nurses was stronger" at places such as HCHP, Kaiser Permanente, and others of similar ilk. Berwick noted that such places "weren't just run by financiers."

"At HCHP Tom Pyle was clearly a businessman, but I think on the whole, Tom had patients' interests in mind. The big players in the Medicare Advantage world—United, Cigna, Aetna, Humana—are financial entities. And they may talk a good game about doing well for the public and for people, but I think if you sit in their boardrooms, you'd find quite a different conversation going on."

In the late 1980s, Berwick departed HCHP and began exploring ways to improve what he considered a broken American health system. In Berwick's view the fee-for-service payment system provided incentives for doctors to focus on the volume of care which was wasteful, inefficient, and too often failed to deliver the quality care patients needed. Berwick's standing grew when he served as a leader of the 1999 and 2001 Institute of Medicine reports *To Err Is Human* and *Crossing the Quality Chasm*. In a deep dive into the issue of safety in American medicine, Berwick and colleagues found a deeply troubling reality: That something on the order of one hundred thousand patients died in US hospitals each year due to preventable medical errors. Never before had there been such an intense spotlight on mistakes in medicine. After Berwick founded the Institute for Healthcare Improvement in 1991, it grew into one of the leading health care quality improvement organizations in the world.

During the early 2000s, a serious re-examination

of health care in the United States gained momentum among clinical practices, large employers, government, think tanks, and universities. Berwick played a role in this period, including in 2008 when two colleagues at IHI (John Whittington and Tom Nolan) defined what they termed the *Triple Aim*, a framework to optimize health system performance, and joined Berwick in putting the ideas into print:[114]

- Improving the patient experience of care (including quality and satisfaction)
- Improving the health of populations
- Reducing the per capita cost of health care

Within two years, the pursuit of the Triple Aim was adopted by many health policy experts as an important framing of what the goals of reform should be.

MOONSHOT

The moonshot that was the Affordable Care Act (ACA)—at 906 pages and 363,086 words—passed the House, without a single Republican vote in its favor, and the Senate, and was signed into law by Obama on March 23, 2010. The ACA provided coverage for an additional 20 million Americans through Medicaid expansion; barred insurers from denying coverage to more than 100 million people with preexisting conditions such as cancer, heart disease, and diabetes; and enabled people up to age twenty-six to be covered by their parents' insurance plans.

The initial roll-out of the exchanges was a mess, with online technical glitches interfering with people trying to sign up for coverage. In retrospect, the chaos of those early days seems eons ago. By 2023, nearly 93 percent of Americans were covered by health insurance and the ACA justifiably claimed its place in history as the third major pillar supporting aging Americans along with the passage of Social Security in 1936 and Medicare in 1964.

PAY FOR VALUE NOT VOLUME

Berwick, Gilfillan, and the authors of the ACA recognized that the financial incentives in health care were backwards, upside down, and sideways all at once. Berwick, Gilfillan, Conway, and others wanted to blow up this system and replace it with programs paying for value (prevention, wellness, an upstream approach to caring for people). The first vehicles for this approach, as prescribed in the act—or, Obamacare as it is commonly known—were Accountable Care Organizations defined by the federal government as "groups of doctors, hospitals, and other health care providers, who come together voluntarily to give coordinated high-quality care to the Medicare patients they serve. Coordinated care helps ensure that patients, especially the chronically ill, get the right care at the right time, with the goal of avoiding unnecessary duplication of services and preventing medical errors."

ACOs fundamentally shifted incentives. Incentives in fee-for-service, Berwick noted, increase the volume of care provided to patients with little evidence that better health outcomes result.

"They are just doing more and more and more," said Berwick. "Elliott Fisher and Jack Wennberg from Dartmouth found in their research" that more was not necessarily better and that often more was worse. ACOs gave providers incentives to invest in staff and change the way they practiced. Essential to the ACO model was upstream preventive care to head off disease in the first place, or—in cases where patients suffered from multiple, complex, chronic conditions—to provide ongoing care to head off deterioration. In 2013, the first year of the program, 1.6 million beneficiaries signed on with 106 ACOs throughout the country.

Berwick took over as Administrator of the Centers of Medicare and Medicare Services in July 2010. Two months later, he convinced Gilfillan to take on the task of setting up and leading the new Medicare Innovation Center created within and funded by the ACA. Berwick thought the Innovation Center "was maybe the most important of all the initiatives under the Affordable Care Act; all the quality initiatives, for sure." Gilfillan, with his medical and business background, was the man he wanted for the job.

The mandate from the ACA was to transform America's delivery system to achieve the Triple Aim—better health, better care, and lower costs. The newly named Center for Medicare and Medicaid Innovation (CMMI) had unprecedented freedom to experiment with new approaches and $10 billion to spend just about any way they wanted at any desired pace over a ten-year period. From the start, the Berwick/Gilfillan plan was simple: Test new models of payment and care delivery, and then measure their efficacy at controlling cost and improving quality.

Additionally, the ACA provided CMS with the ability to expand successful payment and care models, ". . . including implementation on a nationwide basis." The belief then was that this "scaling authority" would allow the Center and CMS to drive major change in national CMS payment policy. From September of 2010, when CMMI was launched, to July 2013 when Gilfillan departed, the Innovation Center launched twenty-nine different initiatives testing various payment and care delivery models. The first of these were directly in support of the CMS ACO initiative.

ACO MODELS FOR EVERYONE

The first strategy, the Medicare Shared Savings Program, aimed to expand and test alternative payment models. It was innovative and proved quite successful. The second strategy, the Pioneer ACO program, in contrast, aimed to do nothing less than to ignite a wildfire of triple-aim focused innovation across the industry and in doing so sending a signal across the medical universe that a new day had arrived; that status quo payment/treatment were targeted for obsolescence. Providers of all types needed to understand the new reality: Transforming the way they practiced under new payment rules was imperative. Payers would have to support the shift to alternative payment models in their own businesses.

Another model known as the Advanced Payment ACO provided much-needed funding up front for physicians to start building their ACOs. The expectation was that hospitals, seeing physician-led ACOs, would have no choice but to build their own. A more edgy approach, as Berwick put it, was the new Pioneer ACO model,[115] which was designed for provider organizations who had broken free of the fee-for-service payment system and gained experience in a value-based payment system where a doctor group took responsibility for not only caring for a population of patients, but also for managing a fixed budget for the care.

Physicians in the Medicare Shared Savings Program (MSSP) ACO were rewarded with bonus payments for providing high-quality care as measured by CMS. In this plan, there was no risk of the physician groups losing money.[116] More-aggressive Pioneer organizations who met quality and cost-savings goals could earn larger bonuses, but there was a catch: In the Pioneer ACOs, provider organizations failing to meet goals risked losing money, which in practice meant having to pay CMS back the advance payments the group had received.

In 2011, CMS received eighty applicants for just thirty-two spots in the Pioneer program. Atrius Health in Massachusetts—a direct descendant of Harvard Community Health Plan where Berwick had worked— was one of the early participants in the Pioneer program eager to try a new approach. (Atrius was later purchased by Optum, Inc., owned by UnitedHealth Group).[117]

Dr. Emily Brower, who led the Atrius Pioneer initiative, told us that she and her colleagues studied data from claims and electronic health records to identify gaps in care and opportunities for improvement. In care for patients with chronic kidney disease, for example, addressing these issues helped the Atrius team diagnose kidney problems sooner and subsequently get patients the treatment they needed sooner.

"The Pioneers were organizations that were experienced in population health management, willing to take some risk and take a flyer on a new program," Dr. Rahul Rajkumar, a senior official at the Innovation Center told us. "These are heroic organizations that were willing to do something bold and take a chance on something. This was challenging at every level. For every one of these organizations, it's like turning around a cruise ship, *changing the mindset of thousands of physicians, getting them to provide care differently.*"

Thirty-two provider organizations in the Pioneer program began with a variety of theories about how to make the model successful. Because saving money on unnecessary care was essential, many of the groups sought to reduce expenses on post-acute care. In practice, this meant selecting people who traditionally might have gone to the highest level of acute rehab and sending them to a skilled-nursing facility (SNF) instead. It meant managing the length of rehab stay. It meant that some people who, in the past, would have gone to a skilled nursing facility would rehab at home. The most-effective way to manage financial risk was by keeping patients healthy enough so they do not require expensive hospitalization. With each passing year, it is clear more and more care for patients can be delivered outside of a hospital setting and this trend aligns with the best interests of ACOs who save money when their patients avoid the hospital, and with patients who vastly prefer home-based care where they recover faster.

"DOCTORS WANT TO DO THE RIGHT THING."

For doctors and their medical teams, changes required to shift from volume to value were daunting. Richard Barasch, whose Collaborative Health Systems operated twenty-five Medicare ACOs, observed that "changing the behavior of doctors from fee-for-service to a value-based environment involves changing in some cases thirty, forty years of behavior and doesn't happen overnight . . . The doctors who embrace it find it very challenging," in part because it requires being more proactive in getting patients in for visits to monitor and control their conditions. This was something new and few doctors' offices had the staff or knowledge to do it effectively. "On top of all of this came a rating system for doctors mandated by Medicare administrators consisting of [33 quality measures.] Under these new Medicare contracts, doctors were rated by patients on a variety of measures, including how well they communicated, whether they used electronic health records, and whether they conducted routine screenings for blood sugar or high blood pressure."[118]

Barasch makes an important observation: "There's another thing going on here, too, and this is sort of interesting from the non-financial . . . viewpoint. The doctors want to do the right thing . . . We're seeing a generational shift in how physicians view their practices . . . They know that pay for performance is coming. Now they are being measured and they want their scores high. They understand that the world is changing and there's a little bit of self-selection in our group with doctors who want to change along with the system. And what we found remarkable in the 2014 reporting period, even the doctors who did not earn bonuses were quite happy with the quality scores that were generated around their practices. They work hard to get their quality scores where they think they should be, and when they're not, the doctors are very, very chagrined."

After the first year of the Pioneers, notwithstanding some turbulence, there was modest good news. Economist Michael Chernew reported that research data showed that "overall the Pioneer program saved a little bit of money for CMS." Economist David Cutler was encouraged by the results "given how historically challenging it has been for physicians to achieve spending reductions in Medicare demonstration projects."

The CMS teams in those early days were thirsting for knowledge, and the knowledge gushed out of a fire hose. "We were all learning as we went," recalls Gilfillan. "It was a whole new world. Pioneer was successful because there were organizations out there who were primed to do this. They were medical groups that were fully capitated for commercial or Medicare product lines. They were successful. These were medical groups around the country chomping at the bit to do this. They knew how to do it. . . . It was successful because we listened, learned, and we had great partners in all these health systems that wanted to do it." (Although in the longer term it would be revealed that Medicare was not paying Pioneer ACOs enough money to expand their growth.)

Pioneer and Medicare Shared Savings Program (MSSP) were sent out into the world to try and answer an essential question: *Could these models of care save money for CMS and also improve the quality of care for patients?* The answer, Gilfillan told us, was yes. ACOs were team-based and it became clear that doctors, nurses, social workers, clinical pharmacists, and others working together could efficiently provide quality care while saving money. Several of the CMMI models for primary care providers demonstrated that PCPs alone were not able to reduce costs, says Gilfillan, because they lacked the full complement of staff and capabilities needed to manage the total cost of care.

The good news was that all three initial ACO models were successful. As of 2023, there are 13 million people involved in ACOs that include more than 500,000 providers. More than 10 million of those people are in the Medicare Shared Savings program. MSSP has consistently produced savings, although not as much as was expected. The Advanced Payment Model established successful physician-led ACOs that produced savings in excess of the investment CMS made. Pioneer ACOs also produced savings in excess of the costs of the program and ultimately was one of the first CMMI models that was certified for scaling by the CMS Actuaries.

CHANGING OF THE GUARD

Berwick was forced out of his role as leader of CMS in December of 2011 due to the unwillingness of Republicans in the Senate to grant him a hearing which could have enabled him to remain in the position.

Gilfillan stayed on for another year and a half, pushing ahead with the new payment programs. After Gilfillan's departure, he was replaced by Patrick Conway who remained in the role leading the Innovation Center until 2017. Previously, Conway had been at Cincinnati Children's leading improvement work and Berwick convinced him to join CMS in June of 2011 as chief medical officer. Thus, when Conway took over the Innovation Center, he had a good deal of CMS experience under his belt.

Berwick and Conway have been center stage in the health care reform movement for many years, and both shared a characteristic that Berwick and others allied with his work saw as essential to changing the system in the right way. This characteristic was what Berwick calls a *"not-for-profit ethos,"* a belief that there was something sacred about the mission to care for people, that it required a selflessness represented by an affiliation with a not-for-profit organization.

Conway followed a similar path. Like Berwick, he trained in pediatrics at Boston Children's Hospital. He served as a White House Fellow, practiced pediatrics at Children's National Hospital in Washington, was chief medical officer at CMS, and later headed the CMS Innovation Center.

"When Don called me," Conway told us, "I was at Cincinnati Children's and planning on being there for a long time. Don said, 'I heard there were two jobs in DC you would come back for, and one was chief medical officer of CMS. Is that true?' And I said, 'that is true.' And he said, 'well, that job's coming open, so would you come to DC and meet with me?'"

Conway flew to Washington, met with Berwick, and accepted the position. At CMS, Conway was also in charge of overseeing quality (though it was not officially part of his title) and, as Berwick put it, "Patrick did good work, absolutely."

One of the other major CMMI models, the Partnership for Patients, launched by CMMI in 2011, aimed to decrease hospital acquired conditions and readmissions. Researchers from the Agency for Healthcare Research and Quality, a government organization, reviewed thirty thousand charts and found significant progress. Partnership for Patients, launched by Berwick and Gilfillan and later overseen by Conway's team, resulted in a 21 percent reduction in harm, or as Conway told us, "125,000 lives saved, $28 billion in cost savings, over 3.1 million infections, adverse events, injuries avoided."

Conway sought to accelerate the pace of change to try and meet the Obama administration's stated goal of having 30 percent of funds in alternative payment models by 2017.

"We went from zero to greater than 30 percent of Medicare payments in less than four years, which was over $200 billion. These are value-based payment models and we had over two hundred thousand signed agreements with providers, health systems, doctors across the nation. I was with Don (Berwick) when he said that it was the biggest shift in Medicare payments ever." The vast majority of that success happened within the ever-enduring Medicare Shared Savings Program, which remains strong more than a decade later.[119]

The bulk of ACOs through the years reached their quality goals and many also saved money for Medicare. Perhaps the most powerful testament to their success is that CMS has established a goal to have all Medicare beneficiaries in ACO relationships by 2030. While Gilfillan, Berwick, and Conway were leaders in these efforts, there were others—hundreds of people—who played important roles in implementing this immensely complex work. Many were permanent employees at CMS who remain to this day.

THE MEDICARE ADVANTAGE AWAKENING

One of the most important things that happened during the CMS experiments with value-based payments was that private companies—insurers for the most part, but also private equity companies eyeing the vast sums of money allocated to geriatric care—saw an opportunity. Silicon Valley was watching. Wall Street was watching. And all could see that a new business had been born— the business of managing the health of older people. They saw the research data from Eric De Jonge and

Tom Edes at the VA, which demonstrated that delivering home-based primary care actually saved money.

ACOs were hard work, and the payoff was downstream—a year, maybe two to earn bonuses. And there was always the chance that, at the end of the day, not only were there no additional payments for physician groups in ACOs, but there were some forced to pay money back to Medicare. Once insurance companies and private equity investors realized that the financial deal with Medicare Advantage was vastly more profitable than ACOs, the rush was on. Portions of the ACO program—with all of its demonstrated improvement and promise—were shoved to the side of the road. CMS was all-in on working through for-profit companies that started Medicare Advantage plans. A major Boston private equity investor called it "the greatest business opportunity in the world."

Did the ACA implementation—all that work by so many people into spreading ACOs—ignite that wildfire of value-based innovation? Yes and no. No, in the sense that the ACO models did not provide a sustainable financial model for health care organizations.

"The financial opportunity did not and still does not generate sufficient returns to justify optimal investing by providers," Gilfillan told us. "Over the subsequent three years it became evident that, while ACOs saved money for CMS, providers did not earn enough through shared savings to make optimal investments. Despite the limited investment and modest returns for providers, ACOs continue to deliver significant savings. Providers continue to work at the model and likely could produce even more savings for CMS if the financial model were improved."

But the answer is also yes. As we noted earlier, there was enough success so that CMS is aiming to have 100 percent of Medicare beneficiaries in ACO programs by 2030.

Gilfillan poses this question: "Was value-based transformation wildfire successful? Yes. How do I know? Because if you ask anybody today to compare the sensibility around, 'Do I need to be delivering value in healthcare,' their answer is 'yes.' We all know that. It is a given. That is a continuous march of models, tests, conversations, publications, and all of that was part of a strategy to build the awareness that we need to think about the total cost of care and people need to be engaging in managing it." America is moving into a post-fee-for-service world. It is a historic step forward.[120]

In our next chapter, we focus on the corporatization of American health care—more doctors now work for for-profit companies than for nonprofits—a reality that would have been unthinkable not so many years ago. As we explore these developments, we also examine the idea from Berwick and many others that doctors and other caregivers who possess the *not-for-profit ethos* are those best suited to lead America's health care system forward, and that, in Berwick's words, those "financially driven entities run by people who've never seen a patient, businesspeople responsible to stockholders . . ." should not be in the driver's seat.

CHAPTER EIGHT

THE MEDICAL-INDUSTRIAL COMPLEX

In 2010, investment in for-profit primary care companies amounted to $15 million. By 2021, it had reached $16 billion.[121]
—Advisory Board, October 3, 2023

SEVEN VOICES

Compassion is the theme that, perhaps more than any other, winds its way through these pages. Compassion and mercy are first cousins, and the definition of mercy that strikes us as most applicable here is the notion from Jesuit theologian James Keenan: "Mercy is the willingness to enter into the chaos of another's life"—into the lives of people who are hurt, confused, sick, afraid, dying.

In many ways, that is what this book is about—people like Rick Gilfillan, Robert Butler, Mary Tinetti,

Eric De Jonge, Patrick Conway, Don Berwick, Sharon Inouye, Susan Reinhard, and others, entering into the chaos of other's lives to ease suffering in a complex and often exasperating health care world. United States Marines are known for many things including their instinct to run *toward* the sound of gunfire. So it is with doctors, nurses, and other caregivers who run toward those whose lives are in chaos. Witness the responses of health professionals during the COVID-19 pandemic. These men and women are guided by a deep well of what Dostoevsky called *the chief and perhaps the only law of all human existence—compassion.*[122]

In this chapter, we step back and listen to the thoughts of seven experts on one of the central questions in health care today: How will we as a society deliver and pay for care needed by an aging population where every day ten thousand people turn age sixty-five? Just as importantly, how will the chosen payment system affect the way older people, especially frail elders, are cared for within the medical system? This question centers on both economics and values.

Many people have opinions on this topic, but we have taken care to select seven people whose knowledge and experience give them a deep understanding of American health care. Included among our seven experts are Dr. Arnold Relman (1923–2014) who achieved iconic status in American medicine as editor of the *New England Journal of Medicine*. We hear as well from three former leaders at CMS including Rick Gilfillan, Don Berwick, and Patrick Conway, as well as from Professor David Meyers at Brown. We hear from Adam Boehler, former head of the CMS Innovation Center and founder of Landmark Health, a home health company, and finally, we hear from Dr. Thomas H. Lee, who practices internal medicine and cardiology at Brigham and Women's Hospital in Boston, teaches at Harvard Medical School, and serves as chief medical officer at Press Ganey, a company that measures patient satisfaction scores for provider groups. Dr. Lee frequently writes articles in leading journals and, appropriately for this chapter, is a member of the editorial Board of the *New England Journal of Medicine*.

Importantly, four of the seven (Gilfillan, Berwick, Conway, and Boehler) have held major leadership roles at CMS where they have had a global view of the US health care payment and delivery system. Five of our seven are physicians (Relman, Gilfillan, Berwick, Conway, and Lee). All seven have worked at government or non-profit organizations while four have experience at for-profit companies (Gilfillan, Conway, Boehler, Lee) and one (Meyers) provides the important perspective of an academic.

"DESTROYING THE TRADITIONS OF MEDICINE."

Dr. Arnold Relman, who passed away at age ninety-one in 2014, was a hurricane of ideas and opinions, a man who despised the possibility that a "medical industrial complex" could impose itself on America. This complex consisted of a "growing network of private corporations engaged in the business of supplying health-care services to patients for a profit—services heretofore provided by nonprofit institutions or individual practitioners."[123]

This, to Relman, was heretical. It was essential, he argued, to "ensure that the medical-industrial complex serves the interests of patients first and its stockholders second"; what he termed "medical entrepreneurship" was doing nothing less than "destroying the traditions of medicine," he wrote in 1991.[124] Relman's standing in the medical world allowed him to get away with opinions that were tough, even harsh. He was a respected researcher, a visiting scientist in biochemistry at Merton College Oxford, and president of the American Federation for Clinical Research, the American Society for Clinical Investigation, and the Association of American Physicians. Most significantly, he served as editor of the *New England Journal of Medicine* from 1977 to 1991. When Relman took over the publication, founded in 1812, he "instilled an editorial purity . . . that eventually became the standard for medical journal publishing," one journalist wrote.[125] The *Boston Globe* noted that during Relman's tenure, NEJM "was the first medical journal to require authors to disclose potential conflicts of interest if they owned stock in biomedical companies when they published research that might benefit those firms."[126]

It was not that Relman considered researchers dishonest. Rather, the editorial purity was one of the many ways Relman fought against what he viewed as the corrosive effects of profit-making in medicine. And so it is that here in the third decade of the twenty-first century American medicine finds itself where Relman feared— where, as noted at the start of the chapter, private investment in primary care has gone from $15 *million* in 2010 to $16 *billion* in 2021.

Relman's warnings may have been heeded to a certain extent while he was alive, but not so much any longer. The medical industrial complex he railed against is now reality. Never before have insurance companies and private equity firms, whose legal obligation is making decisions that financially benefit their investors, been in charge of providing care to so many Americans. Will this trend lead to a brave new world of better, more affordable care? To a place where sick patients are shunned in favor of profit? Or to an amalgam in between? We know that many good things happen in America's not-for-profit legacy health systems. We also know that many good things happen at for-profit companies, as well. This rapid and radical shift toward for-profits delivering care merits serious discussion. In this chapter, we explore the concerns about the new direction as well as opportunities for improvement that it offers.

THE ASCENDANCE OF MEDICARE ADVANTAGE

As of 2023, slightly more than half of retirees sought coverage in a Medicare Advantage program. The speed at which Americans have shifted their loyalty from traditional Medicare to Medicare Advantage, where private corporations manage the plan, is quite remarkable. A generation ago, more than 90 percent of retirees were recipients of traditional Medicare. In historical terms, the change has happened virtually overnight. Yet, for all of Relman's well-intentioned warnings, and notwithstanding the many prominent physicians who share his concerns, Medicare Advantage enjoys high favorability with its beneficiaries.

Some of the developments Relman feared most are playing out on the Medicare Advantage stage. However, if corporate Medicare is as pernicious as Relman believed, why are most retired Americans choosing it over traditional Medicare? Why has the percentage of enrollees in MA climbed from 19 in 2007 to 51 in 2023? Why is it projected to reach 62 percent by 2033?[127] The answers are fairly simple. MA plans have a number of appealing aspects that traditional Medicare does not. For example, according to Medicare.gov, the average cost in 2023 for a Medicare Advantage plan was about $17.60 per month and many lower-income people have a zero-dollar monthly premium. This contrasts with traditional Medicare where costs can range from $160 to $560 per month for part B and \$32–109 for part D, not including the additional costs of a medigap plan.[128]

Most MA plans also offer more coverage for things such as vision and dental. Patients often find enrolling in MA simpler than traditional Medicare. Enrolling in Medicare parts A (hospital coverage), B (medical coverage), and D (prescription drugs), and adding a Medigap protection plan, can create a bureaucratic tangle that's difficult to navigate. In contrast, MA covers in-patient stays, outpatient care, and prescriptions for many older people. Still, there are drawbacks to Medicare Advantage. In return for extra benefits, these plans typically limit patients to specific networks of providers and require prior authorization to see a doctor outside the network.

MA is popular and affordable, but how good is the quality of care? Through his own research, and his exploration of research by others, Professor David Meyers at Brown University told us that based on the beneficiary MA plans appear to have better performance on some outcomes.

"Plans do seem to be pretty effective at keeping people out of the hospital," he said, "and at reducing re-admissions, emergency department [visits], and nursing home use. MA plans also appear to perform well on quality measures such as cancer screenings and disease maintenance. . . . People generally report that they have pretty good quality of care when asked in surveys about the MA program."

The difficulty is that those encouraging results come from studies that were done a decade or two ago, when only about 20 percent of Medicare recipients were enrolled in the MA program. Whether these quality scores have held up over time is uncertain. Meyers cites a 2023 JAMA study which he said found "that ten years ago, if you looked at acute myocardial infarctions outcomes, MA plans were doing much better at keeping people out of the hospital, reducing post-acute care use, and had generally a lot less expensive and higher quality care. However, authors repeated the analysis using more recent data from 2018 and found that almost all those differences have totally gone away as the program has grown. If other research bears this out, I have some concern that many of the quality differences for MA plans might be overstated. There is also concern that part of the reason why it looks like beneficiaries are doing better in MA is explained by plans recruiting healthier people who don't have higher needs . . . We find in a lot of our work that people [. . .] who have the most chronic conditions—the most need—tend to leave the MA program for traditional Medicare."

MA's popularity with retirees is understandable in light of its convenience and extra benefits. The program's popularity with insurance companies and private equity investors is understandable as well: They are given massive infusions of up-front cash—tens of billions annually—to care for people enrolled in their particular program. Three insurance companies together control nearly 60 percent of the MA marketplace—Humana, Aetna/CVS, and the big enchilada UnitedHealth Group. In late 2023, UnitedHealth ranked as the thirteenth most valuable company in the world with a market cap of $499 billion, and annual revenues of $360 billion, a 14 percent annual increase as of the third quarter 2023.[129]

As much as any other company, United saw the potential for MA profitability coming and, in response, built its subsidiary, Optum, seeking to prosper in the space. Perhaps the most relevant statistic concerning United/Optum—and one that would have horrified Dr. Relman—is that United/Optum is the largest single employer of doctors in the nation. Anyone who had suggested such a possibility twenty years ago would have been thought delusional. From 2018 to 2023, Optum doubled in size. The company employs or contracts with about seventy thousand physicians, thirty thousand nurse practitioners, thirty thousand nurses, and five thousand mental health clinicians. Overall, in 2023, Optum served over 130 million people.

PUSHBACK AGAINST DR. WALL STREET

The power of the corporate world to enroll more and more older people into these corporate plans is growing and, as it does so, some fierce opponents of corporatization have inherited Relman's mantle. Here is where there is sharp and sometimes emotional disagreement among clinicians, including among a number of the men and women featured in our book. Gilfillan and Berwick, in particular, have pushed back hard on aspects for the new corporate medical world. As physicians, innovators, former CEOs, and former leaders at CMS, they speak with authority. Individually and jointly, they have such a stature that, when they speak, other health care leaders pause to listen.

In 2021, channeling Relman's spirit, Gilfillan and Berwick authored a scathing indictment of corporate medicine, writing that its for-profit "business model is distorting health care delivery, creating excessive costs for taxpayers and Medicare beneficiaries, draining the Medicare Trust Fund, obstructing the badly needed value transformation of American health care, and diverting the money needed to fund other social services and goods . . ."[130]

Insurance companies are paid a lump sum by Medicare to care for each patient in a Medicare Advantage plan. The amount paid by the government to the insurer depends upon the patient's risk score, which is determined by the number and nature of health conditions afflicting the patient. In other words, the higher a patient's risk score, the more money the insurer gets. Not surprisingly, this has led insurers to scour patient records to push the individual risk score as high as possible. Gilfillan and Berwick refer to this practice as *risk score gaming*, an effort to make patients appear sicker than they might be. They wrote that this "allows the provider firm and plan to harvest a financial windfall just by finding more [billing] codes . . . While the higher risk score might suggest that the population is sicker, that is an illusion created by the risk score game . . . the plan has just collected more codes that make the population look sicker" and thus the MA plan is paid more money by the government.[131]

Meyers has a more nuanced view. He believes that collecting these codes should be a good thing. "Plans should be incentivized to take care of sicker patients," he told us. "The problem is not the idea of risk adjustment, but the amount that plans inflate and invent these codes" and receive more money from the government. "MA plans want people with lots of billable codes, as well as people who are not in need of many different and expensive medical services." At the same time, insurers seek to avoid patients where a code search reveals the need for many expensive medical treatments that could cut into the plan's profits. (The story may well be apocryphal, but a few years ago a rumor went around indicating that MA plans were setting up enrollment locations on the second or third floors of buildings without elevators to make sure applicants were healthy enough to climb a couple of flights of stairs.)

In a style reminiscent of Relman's bare-fisted approach, Gilfillan and Berwick declare that "it is extremely costly to continue to ignore the corrosive, insidious effects of the defective ... risk adjustment system [that] ... is fundamentally redefining our primary care networks, turning PCP practices into insurer-owned or investor-owned coding shops ..."[132] In 2022, the *New York Times* echoed Gilfillan and Berwick in an article entitled "The Cash Monster Was Insatiable: How Insurers Exploited Medicare for Billions."[133] The article reported that "major health insurers exploited the program to inflate their profits by billions of dollars." The *Times* quotes Berwick as saying that "even when they're playing the game legally, we are lining the pockets of very wealthy corporations that are not improving patient care."

For some prominent players in health care, Optum is little more than a rapacious capitalist enterprise. Diane Meier, about whom we wrote earlier in the book, shares the Gilfillan/Berwick perspective.

"Who comes first?" asks Meier. "Is it really the patient? When you're working in a for-profit setting, the boss is the shareholder, not the patient. The most powerful motivating force in health care right now is profit, and patients and physicians are instruments in that money extraction process."

"DO WELL FINANCIALLY AND DO GOOD FOR THE WORLD."

We find quite a different perspective from some of the other voices in this chapter, including Conway, Adam

Boehler (who is now head of Rubicon Founders, a private equity firm based in Nashville), and Dr. Thomas Lee. Conway had served in various positions at CMS for seven years before he was recruited as CEO of Blue Cross Blue Shield of North Carolina. He got off to a fast start, shifting the company to a value-based approach, but left in 2019 and moved on to Optum in 2020.[134] His goal was to shift away from the model where Medicare pays doctors for the volume of care versus paying them for the value of care.

Conway has long believed that "no system or policy or program (traditional Medicare or MA) is perfectly designed. The job of the policymaker is to set up the best incentives possible for better quality and experience and lower total cost of care. The job of organizations in the system is to drive better quality and experience at lower total cost of care." In this respect, Conway explains, his goals have been consistent throughout his career from his roles at Cincinnati Children's to Medicare to Blue Cross of North Carolina and on to United/Optum.

Conway believes Optum can solve a problem long-plaguing medical quality improvement initiatives. For example, while there are many excellent home care programs, most are limited in scope, often severely so, caring for one thousand patients when the need is for twenty times that number. Legacy systems often lack the capital required to scale such programs, but Optum has the resources to invest to scale programs.

"I love scale, and this is the place to drive change and scale. I think it's the biggest private sector opportunity to drive change for a better health system." Conway says that Optum has the capability throughout much of the country to dispatch clinicians to visit frail elders on MA plans in their homes. (He described the programs in more detail in the *New England Journal of Medicine Catalyst* in an article entitled "The Future of Home and Community Care.") Conway has cited the example of a patient who had been hospitalized fourteen times in a year and then went to zero with home care. He emphasizes the benefits of Medicare Advantage, especially to lower income Americans and people of color, including the additional benefits to address social needs.

Boehler is a true believer in paying for value rather than volume. Boehler grew up in a small town outside Albany, New York, where his father was a primary care physician and eventually CEO of a hospital. After graduating from Wharton, Boehler went into the private equity business where he specialized in health care companies.

"There are very few places where you can do well financially and do good for the world, too," he told us. "That's an attractive thing."

In 2013, he founded Landmark Health, a home health company to care for "the sickest of the sick," as he put it, among the older population. Boehler took note of the work that Tom Edes, about whom we wrote in Chapter 2, and his team were doing with home-based care in the VA system and was impressed by the reductions in cost and improvement in quality. Boehler invested a significant amount of time studying the VA model and launched Landmark Health in 2013. The VA model worked because it dispatched doctors and other clinicians to people's homes to provide care when they needed it. Boehler would do something quite similar, but with a private company. The company was successful almost immediately and spread to scores of locations throughout the nation in the ensuing years. In 2021, eight years after starting Landmark, Boehler sold the company to Optum for $3.5 billion.

Why so successful? Boehler says the company aligns the best interests of patients with payment. In traditional Medicare, most doctors are paid for the volume of care they deliver while MA plans pay for the value of care with value defined as keeping patients as healthy as possible by managing their chronic conditions and allowing them to be at home rather than in a hospital.

Boehler cites the hypothetical example of a ninety-four-year-old woman who awakens at 2:00 a.m. on a Sunday with the flu (or other comparable symptoms). A family member may respond by calling 911 and the patient is taken by ambulance to the hospital.

"With a frail elderly person, there is a high probability she will be admitted to the hospital and spend a few days there before going home," says Boehler. "Maybe she catches something in the hospital and later has to be readmitted."

Total cost for the episode, he says: approximately $60,000. But that cost, Boehler argues, is entirely preventable. The fee-for-service payment for a doctor or nurse to visit the patient at 2:00 a.m. might be $120 to $150, he says.

"But if you're going to pay $120 or $150 for that visit, that's not going to happen at 2:00 a.m. on a Sunday." The game-changer is asking "how do you change that incentive so it's about *avoiding* $60,000 in hospital expenses and not about whether you get paid $120 or $150? That is value-based care, which is how do you change incentives to unlock what should happen."

Patrick Conway shares Boehler's view about misaligned payment incentives in fee-for-service health care. "CMS set up ACO programs, while Don, Rick and I were there to shift incentives to quality and experience and lower total cost of care. That has made the Medicare program stronger on the whole and is trying to shift the incentives. MA clearly has incentives for quality, experience, and total cost of care. One can certainly adjust the program over time, but the incentives are clearly directionally correct and aligned with what patients want."

Working in the health care space requires two essential elements, says Boehler. "Number one: You need to do the right thing. People go to medical school and enter into healthcare because they're passionate and they want to do the right thing. The second thing is you have to ensure payment streams. Physicians are not going to do things when there's no payment stream because they can't afford it. If you can reconcile what is the right thing with a payment stream, then you can get action. That is what Landmark does."

We asked Boehler what happens, however, when the financial interests of the company dictate that a company faces declining demand and must reduce expenses? "In outcome-based care," he says, "you have incentives to lower cost and improve quality, and people will say, 'does that mean you can shunt care?' The truth is, in certain circumstances that could happen. I think it's few and far between because shunting is pretty bad to do and you're getting into criminal activity if you're actually aggressively shunting. You need to put the right monitors in place because no system is perfect. I overwhelmingly prefer the incentives of an outcome-based system, but no system is perfect. The best thing we can do is try to align incentives for the outcomes we want." The government's job, he adds, is to "ensure that patient quality is set up the right way and to set those guard rails."

AMONG THE STARS

Among the more controversial aspects of MA are the star ratings used to measure quality. Optum and other companies boast about their ratings, but Meyers urges caution. In all, about thirty-five to forty different measures are mixed together to calculate a star rating. Gilfillan and Berwick have observed that from 2017 to 2022 the number of MA plans with star ratings above 3.5 "increased from 49 percent to 68 percent and membership in the 'above average plans' went from 68 percent to 90 percent." They have referred to this phenomenon as the "Lake Wobegon effect," from Garrison Keillor's show "A Prairie Home Companion" where everyone is above average.[135]

Meyers says that the research into the star ratings backs up that skeptical view. "Star ratings are not predictive of better outcomes," he told us. "After controlling for beneficiary characteristics such as health and income, research has found that enrolling in a higher rated plan brings no significant benefits for beneficiaries."

It gets worse, he says. In a 2023 research paper, Meyers found that "given that much of the growth in Medicare Advantage has taken place among racial and ethnic minority populations, understanding the equity of care provided in Medicare Advantage is also important. Prior work has found that Black and Hispanic beneficiaries tend to be enrolled in lower-quality plans, as assessed by a variety of metrics including star ratings, physician network breadth, and the likelihood of the plan being terminated from the program."[136]

"IT IS THE ETHOS THAT MATTERS."

In addition to working at Press Ganey and teaching at Harvard's schools of medicine and public health, Tom Lee practices internal medicine and cardiology at Brigham & Women's Hospital in Boston. Lee grew up in a small town outside Philadelphia in a family that was gifted intellectually. Tom and both of his brothers went to Harvard as undergraduates. Years before that, however, Tom Lee had an experience seared into his memory when his mother took Tom and his brothers to Taiwan to visit family for a humid summer when Taiwan was not nearly as advanced as it is today. There were impoverished people in obviously difficult circumstances. Much worse, however, was when Tom went to visit his mother's brother in a mental institution. He was confined there for a condition Tom did not understand at the time. What he saw and understood, at least to an extent, was "the suffering of the people around me in the hospital," he said. "It was so obvious."

Lee has spent his professional life working to alleviate suffering. He worked for many years in both for-profit organizations and not-for-profit organizations, and he resists weighing in on the merits of one side or the other.

"I think virtually everyone in healthcare does good things some of the time and bad things some of the time," he told us. "The goal is to try to maximize the circumstances in which they do good things and minimize the circumstances where they can do bad things. Quite frankly, nonprofits act pretty much the same as for-profits in what they do for the most part, absent not paying taxes."

This is heresy to many people in the traditional not-for-profit side of health care. Theirs is a mission focused on people, not money. That is true with many doctors, nurses, and other caregivers. Many health systems and physician practices also possess a mission-based culture. There are differences, of course, and they are important. As a colleague of ours put it: "In stark terms, our mission is people while the corporation's is profit. Does that mean they put out a bad product? Not at all. But . . . they don't do the things we do: fund and conduct research, fund and do medical education, and fund and do community outreach."

David Meyers, the Brown University Medicare Advantage expert, takes the position that "it is the ethos here that matters, not the non-profit status itself. There is a lot of health policy literature that finds many non-profit health systems don't really operate any differently than for-profit entities. There are plenty of non-profit actors in the healthcare sector that are *very* profit-driven. While their profits might not go to shareholders, they do enrich the companies. There are many insurance plans that are 'non-profit' that operate just as cutthroat as the for-profit entities. The ethos and culture of an organization can vary, but tax status itself is not a very useful distinction."

Don Berwick recalls an incident from his days at Harvard Community Health Plan in the 1980s that emphasizes Meyers's point. "I don't think HCHP was completely pure," he told us. Berwick recalls a meeting of the organization's executives when it was revealed that "sometimes patients died, but their employers, who were paying the Harvard Plan for their care, didn't know they were dead so they continued to pay for months, or even years, afterward. For people who had died. But because Harvard Community Health Plan had medical records, it knew who had died. And there were not a small number of people for whom they would continue to collect money who were no longer alive. And I remember raising my hand in the room and saying, 'wait a minute. We are knowingly taking premiums for the care of people who are dead. We know they're dead and the payer/employer doesn't and we keep the money?' And everybody looked at me like I was naïve. *Of course, we keep the money.* There was no question in their minds. And I remember [the CEO] turning to me and saying, 'okay, we'll return the money. You want me to take it out of your budget?' That was his response to me."

GAME-PLAYING VS. VALUE CREATION

"I think Don and Rick are right," Tom Lee told us. "There is a lot of greed going on . . . and there is a lot of game playing as opposed to value creation. I understand why people are mad at Optum and United, but they actually do a lot of really good things, too. . . . And people in the not-for-profit sectors do things which are not really aimed at improving health outcomes with efficiency. They're doing things to optimize their financial situation, too. I tell my colleagues, 'think about two different types of activities.' There is *gameplaying* and there is *value creation*. Gameplaying is where you're trying to get paid as much as you can be paid for what you're doing and there is nothing dishonorable about that. It is a completely natural and normal thing to do, and I always want people who are good at gameplaying on my side. I think it is naïve to say we need a world where there isn't going to be gameplaying."

The value creation side of the ledger, according to Lee, is where improvements are made in outcomes and efficiency.

"I tell doctors, 'you actually have very little to add to the gameplaying side of things. On the other hand, you have a tremendous amount to offer in the value creation side where you actually redesign things; that is where you're going to do the most things to make you proud and actually make healthcare better."

Meyers shares much of Lee's thinking but he also cautions that "the job of policymakers and researchers is to ensure that the [gameplaying] and value creation remains in the best interest of patients while at the same time not bankrupting the Medicare program."

THE FUTURE?

Both Meyers and Lee see MA as a powerful force in the future of care delivery in the United States: "It's clear from experience that fee-for-service doesn't work, doesn't incentivize the best outcomes," Meyers told us. "MA has more potential to manage patients' care and deliver on outcomes because the incentives are aligned. The more innovative players are just raising the bars for what is possible when you break free from the fee-for-service system."

Meyers sees MA growing significantly in years to come. Just over half of Medicare beneficiaries are already in MA and "we can probably expect getting to 70 or 80 percent MA over the next fifteen to twenty years because the price of these plans, the additional benefits that they're able to offer, and how profitable these plans are, provides a strong incentive for these plans to grow. At that point, substantially curbing a program upon which the great majority of older Americans rely would be unrealistic. The MA train is too big and moving too quickly."

Meyers believes that a potential long-term benefit of Medicare Advantage could be that CMS does not have to manage the care for all beneficiaries. "In traditional Medicare, if there is any care management, it's CMS's responsibility, although they don't do that much," he says. "Whereas in Medicare Advantage, you're essentially outsourcing managing care. You're outsourcing all the billing and care coordination with the hope that the payment incentives that [insurance] plans have will incentivize them to ensure patients are getting the care they need. The other argument that might be made about Medicare Advantage being a larger fraction of a program is that [in MA] CMS chooses how much to pay these plans each year. Theoretically, it's a way that you can cap your total spending in the Medicare program because everybody's enrolled in Medicare Advantage. CMS could then say, 'this is how much we're paying the plans to take care of people.' Whereas in a fee-for-service model—traditional Medicare—CMS is on the line for every service that needs to get paid." Although the reality, Meyers notes, is that CMS has no history of cutting back on payments to plans.

Meyers envisions a future where Medicare will probably be some MA model with maybe 20 percent of people leftover in traditional Medicare as a public option that has less restrictions, that relies on value-based services like Accountable Care Organizations. Potentially, that would include the sickest 20 percent of Medicare patients "who aren't able to deal with the care restrictions that the plans put into place. A fraction of all Medicare recipients are the sickest part of the population. The other fraction's probably the wealthiest part of the population because if you're wealthy and you don't need to worry about how much you're spending on healthcare, then traditional Medicare is great because you can see any doctor. If you can afford those premiums and afford the Medigap plan, it can be a pretty good option. But it'll probably be a mix of people that can afford everything easily and then the sickest people who find the care restrictions in MA detrimental to their access to care."

Says Lee: "The potential is there for real redesign in care, real value creation that simply is not there in fee-for-service Medicare where the focus is purely on volume. I do believe that having more and more people in Medicare Advantage is a good thing. It is not the final destination, but at the very least it is a step away from rewarding volume as opposed to value. I think we are just at the beginning of that process."

Lee notes that many legacy health systems are stuck in traditional Medicare and unable to make the leap from fee-for-service to value-based care. "What does it mean to really organize around the needs of patients?" he asks. "As opposed to what doctors do in maximizing their income as they do the good things that doctors do? Here is a contrast that I'll draw for you. At Devoted Health, which is a Medicare Advantage Plan, they believe in basically low-tech blood pressure monitors, bathroom scales, not super advanced stuff. But they organize people around the low tech so that when a Devoted patient is diagnosed with hypertension, they've got a blood pressure monitor, which is uploading their blood pressures, and they have teams of people who are adjusting the medications every two or three days."

Lee contrasts that with his own clinical practice in Boston where, when he sees a patient, he makes whatever adjustments in medications are required and then has a return visit with the patient in two to three months when he has his next opening. He likes the Devoted approach where under MA contracts the clinicians have an incentive to constantly monitor patients and make whatever adjustments (medication or otherwise) are needed to keep the patient healthy and out of the hospital.

But the organization where he works—Mass General Brigham—is "not set up for that. The Devoted patients who are getting their blood pressure managed by a team organized around patients getting their blood pressures controlled much more quickly and much more efficiently and better controlled than the patients who see a pretty good doctor like me in the traditional model. And that is because everything that happens in the traditional model is structured around when the doctor has time that will get reimbursed. That is just one small example of the very marked contrast with what is possible when you're not basing everything upon what doctors are paid for on a fee-for-service basis. Not many Medicare Advantage Plans have gone to that step Devoted has in their chronic disease management, but they're all going to get there. Over the next ten years, they are all going to get there."

Will the shift toward Medicare Advantage be in the

best interests of patients? Will insurers make enough money to stay in the business long-term? Will the American people come to the realization at some point that corporate medicine is now woven into the fabric of our care delivery system? Will they care? What is clear, observes Meyers, is that the corporatization and financialization of medicine is here, and those of us in the business of providing care must be ever more vigilant to see that the best interests of patients are kept front and center.

CHAPTER NINE

THE UNDISPUTED WORKHORSES
OF AMERICAN MEDICINE

"Crisis" is a word too frequently tossed around in health care, but when it is used to describe the current mayhem with mental health and primary care, it may not be an overstatement. In fact, the need to improve access to both mental health and primary care is urgent, bordering on an emergency, especially for older adults with multiple chronic conditions.

Primary care and mental health care once occupied separate galaxies in the medical universe. No longer. In recent years, more and more responsibility for mental health care has forced itself into the house of primary care, to the point where they are now cohabitating. It was never intended that responsibility for treating mental health would fall to primary care physicians, but necessity in the medical world is an irresistible force.

Demand for emotional health services is rising, even

as the supply of trained therapists declines. More than 50 million Americans experienced mental illness in recent years, yet most (28 million) received no treatment due largely to staffing shortages.[137] Older people have a significant vulnerability to depression and anxiety, as they experience declines in physical and cognitive abilities.[138] Adults over sixty-five with serious illnesses such as cancer, stroke, or dementia suffer from significant rates of depression ranging from 30 percent and higher.[139]

It is important not to overstate the problem. In fact, the great majority of older patients have a generally positive view of life and thoroughly enjoy their families as well as a wide variety of hobbies, sports, reading, engaging socially, and much more. Patients who are exceptions often struggle with the most common mental health issues, depression and anxiety.

Estimates of older adults suffering from mental health vary widely. This is particularly true in light of data gathered during the pandemic when life conditions were especially difficult for many older people. A Kaiser Family Foundation report stating that up to 25 percent of adults sixty-five and older experience anxiety or depression was based on research in 2020 in the depths of the pandemic. This would suggest that 20 million older Americans suffer from mental illness.

Whatever the precise number if we look beyond the pandemic it is clear that many millions of older Americans are suffering. Consider life experiences of men and women in their seventies, eighties, and beyond. Painfully few human beings escape life without reversals—professional or personal. Those setbacks exact a toll. Divorce, for example, leaves some older adults feeling abandoned, their self-esteem damaged. Strained relationships with family members, unfortunately all too common, are often draining and quite painful. Some of the worst tragedies—the loss of a child or spouse—can send otherwise stable people spinning off into dark places. Add to these the loss of friends as we age.

We have treated older people who retire from jobs they enjoyed and find themselves rudderless, without a focus or purpose in life. Others feel debilitated by their loss of income. And though many older people do not say it out loud, they feel targeted by ageism and the implication that without steady work and income, they have little value to society any longer. The pandemic exacerbated the problem as loneliness and isolation cut off older adults from many of their sources of joy in life. On top of all this are the primary reasons we see older patients—physical ailments, often quite painful, that restrict or prevent a person's ability to do the things that bring joy and satisfaction. Such losses can have profound effects on a person's mental status.

The encouraging news is that the old status quo of treating a particular physical malady and sending the patient on his or her way, is, increasingly, a thing of the past. Doctors and other caregivers have become skilled at identifying signs of mental health disturbances. Primary care teams, in particular, have developed advanced skills for both identification and treatment of such conditions. The evolution of primary care teams dealing with patients' mental health issues is a work in progress. Here again is a supply/demand problem: While demand for services from primary care doctors is rising, the supply is declining. Just as fewer medical students choose to specialize in psychiatry, so, too, are fewer students choosing primary care specialties (including internal medicine, geriatrics, family medicine, and pediatrics). Students frequently hear about the long days, late-night calls, and therefore opt for positions with shift work and scheduled off-time.

The mental health problem in the United States afflicts people in every walk of life, including the very same doctors called upon to care for depressed patients. The *New York Times* reported in 2022 that "nearly two-thirds of doctors are experiencing at least one symptom of burnout . . . defined by increased emotional exhaustion, a more distant approach to the job, and a declining sense of personal accomplishment."[140] The pandemic pushed these numbers into the stratosphere, but burnout was already on the rise pre-COVID.

In 2019, research from the National Academy of Medicine found that "many doctors' dissatisfaction with their work could be caused by an incongruence between what they cared about and what they were incentivized to do by the health care system."[141]

This is worth pondering. How can it be that doctors—who are trained to heal—are financially incentivized to do something other than what they believe is in the best interest of their patient? Here is where the fee-for-service payment system, which works well for certain aspects of care, reveals its fatal weakness.

A 2020 report in the *Journal of the American Medical Association* found that 44 percent of doctors "experienced symptoms of burnout, characterized by emotional exhaustion and/or depersonalization, at least weekly." The JAMA paper was revealing for another reason: It showed that physicians have "significantly higher resilience scores than the general employed US population" with resilience defined as "the collection of personal qualities that enable a person to adapt well and even thrive in the face of adversity and stress."[142] Clearly a strength. What is striking, however, is the finding that "even the most resilient physicians had substantial rates of burnout." In other words, these doctors experience emotional exhaustion often accompanied by anxiety and sometimes depression and *still keep marching forward caring for patients!* The article suggests that targeting "characteristics of the practice and external environments (e.g., regulatory requirements) that contribute to burnout" could help. "For example, targets for improvement include inefficient workplace processes, excessive workloads, and negative leadership behaviors."[143]

Physicians are not shy about telling anyone within

earshot that time pressures, chaos within the system, the power of insurance overlords, and much more make life stressful and sometimes infuriating. Doctors "fare better in organizations where they are not compensated for individual productivity, are not under time stress, have more control over clinical issues, and are able to balance family life with their work," according to a report from the Agency for Healthcare Research and Quality.

Perhaps the most important recommendation from the AHRQ authors called for "reducing the physician panel size to 1,800 patients, increasing flexibility for longer patient visits, reducing the number of face-to-face visits per day, and increasing care team staffing improved work satisfaction and burnout rates."[144] However, this desire, on the part of doctors, collides head-on with productivity demands from payers and leaders of provider groups.

As is so often the case in health care—along with the complex and concerning facets—comes the uplifting part. Here is where we find the resilience of primary care doctors who are taking on the gargantuan task of directing the nation's mental health traffic. The pandemic demonstrated that health care workers, when faced with a devastating crisis, adapted well to the new reality. We have seen firsthand over the past several decades just how resilient people in health care can be, perhaps especially primary care doctors, who have become the undisputed workhorses of American medicine. Primary care physicians routinely screen patients for mental health concerns and, while not trained in psychiatry, are often able to treat mild to moderate mental health issues themselves or in concert with specialist team members.

Physicians in primary care don't get nearly enough credit for the work they do. In recent decades, the care burden has shifted to placing disproportionate weight on primary care. The result is complex and concerning. Complex because they are asked to do more work with less time, concerning because something has got to give. Too many doctors are being asked to do too much too often without the proper team support they need and without the time off they require to clear their minds and achieve a peaceful steady state. A National Academies on Science, Engineering, and Medicine report in 2021 defined the problem:

> High-quality primary care is the foundation of the health care system. It provides continuous, person-centered, relationship-based care that considers the needs and preferences of individuals, families, and communities. Without access to high quality primary care, minor health problems can spiral into chronic diseases, chronic disease management becomes difficult and uncoordinated, visits to emergency departments increase, preventive care lags, and health care spending soars to unsustainable levels.[145]

INCREASING ACCESS, COLLABORATIVE CARE

We asked Dr. John Q. Young, System Chair and SVP for Department of Psychiatry at Northwell and Zucker School of Medicine at Hofstra/Northwell, how primary care doctors, without enough time or training, are able to shoulder so much of the mental health burden? Young has a fascinating history, not unlike some of the other physicians we write about. After graduating from Harvard with a joint degree in social studies and comparative study of religion, he attended medical school at the University of California San Francisco and considered specializing in internal medicine and primary care until he did a rotation in psychiatry.

"I was just really drawn to folks who were suffering with mental illness," he told us. "There are a lot of important questions that come up—*why we do what we do? Where does meaning and joy come from? How do you help people who are psychotic or really depressed?* There's something really compelling about it."

Young was on a mission of mercy from a young age. Prior to medical school, he worked in South Africa for a church-based human rights and anti-apartheid organization founded by Desmond Tutu. He worked another summer in a slum in India with an NGO started by an Asian/Indian liberation theologian and worked on a mobile clinic during an outbreak of cholera.

Young told us the key to treating as many patients as possible is for doctors in all specialties to collaborate with trained counselors. This collaborative care model has gained traction throughout much of the nation and generally works pretty well.

"We will never have enough mental health clinics to meet the need," Young explained. He stressed that we must integrate mental health into the medical ecosystem with primary care doctors as well as specialists in a wide variety of categories, from oncology to cardiology and beyond.

With its roots at the University of Washington, this collaborative care model has existed for about twenty years, but has taken off in the past decade. The system integrates trained counselors who assist the primary care physician in treating a variety of mental health problems. The counselors, in turn, are backed up by a psychiatrist who serves as a consultant to the whole team—guiding counselors in how best to handle different types of patients. The psychiatrist is also on standby to make judgments about patients who may need more intensive therapy. In some cases, the psychiatric lead on the team works in the primary care office; in other cases, they are available by phone or virtually.

Essential to the model's success is having a strong mental health care manager as the team's pivot person. This role can be filled by nurses, social workers, psychologists, or licensed mental health counselors. The approach is a force multiplier, where the psychiatrist's role is much different from the norm. A psychiatrist providing face-to-face therapy in an office might see ten patients per day, while a psychiatrist integrated into a collaborative care team can work with a care manager and primary care physician to manage treatment for ten times that many patients.

"The model works pretty well for mild to moderate psychiatric comorbidities in a medical practice," Dr. Blaine Greenwald, Vice Chair for Department of Psychiatry and Division Chief for Geriatric Psychiatry at Northwell, told us, but there are limits. "When it gets severe, the person is suicidal, the person is delusional, then it overwhelms the primary care practice to be able to handle that. In such cases, patients are referred to our psychiatric clinic where they are cared for by a psychiatrist."

In an ideal world, perhaps every patient would have direct access to a professional trained in mental health, but in today's reality we embrace mental health as an integral part of primary care. Treating the mind and body is the surest way to health. Until recently, we have lived in a split world, with physical health on one side and mental on the other. This division has long been true for health systems, educational curriculum, and our public discourse. But embracing both is what works best for patients.

In doing so, we have found that our primary care doctors, with more and more experience treating patients with mental health issues, have become quite skilled in this complex discipline despite their lack of formal training. This adaptation is a wonderful testament to the intelligence, flexibility, and commitment of primary care doctors.

FAITH-BASED TRAINING IN MENTAL HEALTH

Another promising novel way to provide care for people in the community with mild to moderate depression and anxiety is through faith groups. Older people disproportionately attend worship services and when they do, they feel they are in a trusting place where some are willing or even eager to discuss their health concerns. Young recently received a grant to embed a full-time therapist in three different faith-based organizations, and then to train members of the organization, laypeople and laycounselors, equipping them with skills to care for people in their community.

"We're training Muslim, Jewish, Christian leaders in psychological first aid to help them provide support to members of their parish or their congregation or temple," Young told us.

This approach stems from Young's experience in global health where he worked among the poor where resources were scarce.

"It's what we've learned in the global south, from folks in extremely under-resourced, high need areas like India, parts of Africa, central South America," he told us. "Maybe you have a high school community health worker that's serving eleven villages in South India. We've been doing this in global mental health work for a couple decades and we've shown that you can train lay people to provide empirically supported psychological intervention."

Young is passionate about this work and cautions that he does not want "to oversell this" in part because it has minimal penetration in the United States so far.

"I think this has a lot of potential and there's a bunch of us in different parts of the country who are really excited about this." Enthusiasm has grown in the UK as well, where thousands of people are being trained to do this sort of work.

The COVID-19 pandemic also demonstrated the power of digital connections for all patients, including older people. Through the use of telehealth and virtual technology, there are many new apps claiming to help people with mental health. Young told us that embedding Cognitive Behavioral Therapy for insomnia within mobile apps helps older adults with the common problem of sleep deprivation. Young also says that there is an additional benefit when cognitive therapy is used in combination with medication.

Certainly, telehealth has been a great help for older people with difficulty leaving the home. We have also found telehealth especially effective in emergency situations where someone has suffered a mental issue and appeared in one of our emergency departments.

The ability to have a psychiatrist on the screen from a remote location, usually in a matter of minutes, can be extremely beneficial for a patient in distress. This meets the goals of trying to reach patients where they are—the right care in the right place at the right time.

ROBUST PRIMARY CARE: FOUNDATION OF HIGH-QUALITY HEALTH SYSTEM

Depending upon one's point of view, the condition of primary care and mental health services in the United States is either a wonderful opportunity or a woeful problem. Perhaps both. We are optimists. So many people and organizations are working on these two issues that we believe there are brighter days ahead in both areas. However, that day is not tomorrow, and it will not come easily.

Dr. Christine Ritchie, Director of Research for the Division of Palliative Care and Geriatric Medicine at Massachusetts General Hospital, says she believes that mental health capability "needs to be baked into every single aspect of our work from a preventive perspective all the way to a crisis management perspective . . . There is so much good evidence out there for strategies to improve people's behavioral well-being, and it's just not been integrated into our training."

A bright spot is that there is now widespread awareness of the problem and multiple efforts to find solutions, including bringing psychiatry in from the cold and integrating mental health capabilities into primary care settings. Some research finds the use of Artificial Intelligence (AI) helpful, while many health systems are utilizing virtual visits that put the psychiatrist or psychologist in front of the patient virtually within a matter of minutes. These are positive steps, as is the increased attention to the issue which Dr. Harold Pincus, Professor in Department of Psychiatry at Columbia University, says is partly due to the issues having "been ignored for so long. It's like a dam beginning to burst." Another silver lining is that proposed Medicare policies have the potential to increase payments for crisis care, substance use disorder treatment, and other complex patients.

Dr. Robert Valdez, Director of AHRQ, and Dr. Arlene Bierman, Chief Strategy Officer for AHRQ, emphasized the challenges posed by the fragmented health care system in the United States and the difficulty of navigating it, along with the burden its complexity places on patients and families. Many older adults work with multiple clinicians who do not communicate with one another and who are unaware of the full breadth of care being provided to the individual patient. The lack of coordination among clinicians compounds the negative experiences that many patients have when encountering clinicians who have little understanding of their needs, desires, and quality of life.[146]

Clinicians and insurers are sounding the primary care alarm. During a conference at the Leonard Davis

Institute of Health Economics at Penn in 2022, a damning consensus emerged that primary care in the United States "is fragmented, underfunded, and hobbled by widespread inequities, chronic workforce shortages, cost and payment barriers, uneven distribution of providers, a long unwillingness to acknowledge and address the profound health impacts of social determinants of health, and a medical education system that holds primary care in low esteem." One of the conference participants, Linda McCauley, PhD, RN, Dean of the Emory School of Nursing, went so far as to say that the foundation of primary care is "crumbling." In the face of this disturbing analysis, the conference leader, Rachel M. Werner, MD, PhD, director of the Davis Institute, made a statement that resonates today and will do so for years to come: "High-quality primary care is the foundation of a high-functioning health system, and most evidence suggests that primary care is the only medical specialty where a greater supply produces improvements in population health, longer lives, and greater health equity at a lower cost."[147]

Dr. Butler would no doubt be encouraged that much of the suffering of older people no longer hides in shadows. The aging revolution has shed bright light on the lives of older adults in their homes, in the community, in nursing homes, and in hospitals. This is an enormous achievement.

Shortly before he died, Butler published his final

book, *The Longevity Prescription,* in 2010. His aim was to use the accumulated knowledge and resources of the International Longevity Center, which he founded in 1990, to offer the best strategies for living longer and well. As Achenbaum notes in his biography of Butler, the book promised neither magic pills nor elixirs but nine suggestions, five of which involved positive outlooks and connections with others: (1) maintain mental vitality; (2) nurture your relationships; (3) sleep well; (4) set stress aside; (5) connect with your community; (6) live an active life; (7) eat your way to health; (8) practice prevention; (9) stay with these strategies.[148]

Finally, Butler stressed the importance "to do things that rouse the quiet stream of happiness that you know is there."

CHAPTER TEN

THE POWER OF SMALL THINGS

A GIFT FROM DR. FARMER

Paul Farmer, MD, PhD, a physician and anthropologist, was one of those rare individuals who sets out to change the world and actually does so. Not the entire world, of course, but those desperate places where the poorest gasp for breath. Farmer was a Harvard-trained physician in Boston, but his patients did not come to his ultra-modern space in the heart of the city's medical area. Instead, Farmer went to his patients—travelling to Haiti, Rwanda, Lesotho, Liberia, Russia, the Navajo Nation—wherever the world's most vulnerable needed him.

In 1987, when he was just twenty-eight years old, Farmer and his colleagues founded *Partners in Health*, an organization providing medical care to the poor.[149] He was a charismatic individual, a master fundraiser who attracted generous donations to support his efforts. Later, he achieved the distinguished title of University Professor at Harvard while serving as chair of global health at Harvard Medical School and chief of the Division of Global Health Equity at Brigham and Women's Hospital in Boston. He also spent time in his research laboratory, where he created new drug strategies that cured tuberculosis among many people, most in poor nations.

Farmer was innovative in his approach, partnering with community health workers with minimal education to deliver care in remote corners of the world. He provided free care in places where electric power, water, and basic medical technology were sometimes unavailable and often unreliable. As if these accomplishments were not enough, he won a MacArthur Fellowship and was described in the book *Mountains Beyond Mountains* as "the man who would cure the world."[150]

On February 21, 2022, Farmer was in Rwanda to minister to his patients when he died in his sleep, felled by a cardiac event. He was sixty-two years old.[151]

Farmer left a rich legacy, part of which is a gift he bequeathed to those of us who occupy a place in the world's vast and tangled health care universe. Although it is not a tangible gift, it is nonetheless a precious one directly relevant to this book. His gift is the idea of what he called "*accompaniment.*" The term comes from community health workers in Haiti—called "*accompagnateurs*"—who *accompany* patients in their health journey. To accompany someone, Farmer told the graduating class at the Harvard Kennedy School in 2011, is "to be present on a journey with a beginning and an end . . . There's an element of mystery and openness . . . *I'll share your fate for a while, and by 'a while' I don't mean 'a little while.'* Accompaniment is much more often about sticking with a task until it's deemed completed by the person or person being accompanied . . . [it] requires cooperation, openness, and teamwork . . ."[152]

PATHWAY FORWARD

Modern medicine exists within two different spheres. One contains the really big things: AI; new drugs for cancer, diabetes, obesity; imaging technologies more precise than ever; breakthrough ideas for payment reform; organ transplants; and more. These are the subjects of page one articles in the *New York Times*, segments on *60 Minutes*—glory, recognition, and money.

In contrast, the other medical sphere is small. It is humble, quiet, and, perhaps above all, intimate. It does not require Nobel breakthroughs or space-age technology. No *60 Minutes* here. But the small sphere contains the nucleus of the health care atom—a place where doctor and patient come face to face in encounters that can feel nearly sacred. These encounters are honest, even raw at times, when a patient's primal fears tumble out in a sort of frenzy. This is the place where the most promising medical pathways will lead.

If there is a rough blueprint for where health care

should be headed, it comprises the ideas and work of the people about whom we write in these pages. Take the work of Eric De Jonge and other home-based primary caregivers, for example. Unfortunately, Congress allowed the Independence at Home program to sunset at the end of 2023. In light of the data underscoring the program's success, this made little sense. Home-based primary care is embraced by patients and providers. In many cases, it saved money for Medicare. A continuation of the program seems an obvious choice, particularly with America's rapidly growing population of frail elders. An expanded, sustained Independence at Home program is something patients and providers want, and Medicare should embrace.

The work of Mary Tinetti and Sharon Inouye on two of the toughest geriatric syndromes—falls and delirium—demonstrated the efficacy of shining a light on such conditions. Tinetti sounded a worldwide alarm with her research on falls, while Inouye formulated ways to not only detect delirium, but to manage and even prevent it as well. Further attention to both of those issues is warranted, particularly research that points to more-effective fall prevention initiatives. Expanding (or even mandating) the application of Inouye's techniques for diagnosing and preventing delirium (such as the HELP program) will continue to save many lives.

Over the course of many years, Diane Meier,

Rosemary Gibson, and the Robert Wood Johnson Foundation raised awareness of palliative care, which has now been embraced by nearly every major health system in the nation. Susan Reinhard at AARP along with other advocates for family-caregivers have alerted the nation to the plight of 40 million Americans caring for ailing family members. The burdens these people face daily are finally out of the shadows and into the light of day, where virtually the entire nation recognizes the urgency of the situation. Their efforts were aided by Adam Boehler, who developed novel models of care to support frail elders at home. When it comes to the business of health care, innovators such as Don Berwick, Rick Gilfillan, and Patrick Conway have altered payment models to reward performance, improve quality of care, and strengthen efficiency.

PATIENT PRIORITIES CARE

Looking to the future, we are attracted by the age-friendly approach and the idea of *what matters* and how it can be a framework to fight ageism and promote person-centered care. We also recognize its shortcoming on a practical level. We need to take more-aggressive steps to make age-friendly care the nucleus of the doctor-patient relationship. Tinetti and colleagues have articulated an idea called Patient Priorities Care, which exists within medicine's smaller sphere and embodies much of the thinking and ethos of what this book is about.

Most importantly, Patient Priorities Care adds muscle and structure to the age-friendly approach.

Tinetti and colleagues officially launched the Patient Priorities Care initiative in 2014, but, in a way, she started delivering it forty years earlier on that snowy night in Rochester where she found her patient Alma Davis had fallen and been living on the floor for weeks. Patient Priorities Care inherits the wisdom of *what matters*, then takes the next logical step: Turning what the patient most wants into reality. Tinetti wanted to construct an approach for how clinicians might determine what matters to each patient and then *what to do about it*. She says that many physicians recognize that what matters is the heart of the age-friendly 4Ms, but they don't know how to integrate the patient's wishes into their practice. Tinetti set out to change that.

Starting in 2014, she and her team spent more than a year talking with patients, caregivers, primary and specialty physicians, health system administrators, nurses, physical therapists, clinical pharmacists, and social workers to identify an effective approach to decision-making for older adults with multiple chronic conditions.

"The concept of Patient Priorities Care came from people who have been practicing in the care of older adults for decades," she told us.

The traditional status quo of treating each disease is often burdensome for patients with multiple conditions.

"How do you define outcomes in somebody who has five conditions?" she asks. "How do you compare preventing a stroke from preventing a hip fracture? People vary in what they most want from their health care. If you're not sure, what else would you align care with except for each individual's health priority? That is, with the outcome each person you prioritize in the face of tradeoffs."

To translate what matters into decisions, she and her team knew that clinicians have to ask specific questions: What are the patient's "actionable, realistic, specific health outcome goals?" For each individual patient, what are his or her preferences for tradeoffs, recognizing that some interventions can be helpful in some ways but harmful in other ways? And, finally, what was the *one thing* that is most important to that patient's health?

Fortuitously, Amy Berman from the John A. Hartford Foundation was interested in supporting work addressing the interaction between primary and specialty clinicians. The foundation along with the Patient Centered Outcome Research Institute provided financial support to train physicians and other caregivers on how to communicate with patients to identify that *one thing*.

Part of the challenge is that physicians make assumptions about what patients most want and those assumptions can often be wrong. Dr. Barry Wu, professor at the Yale School of Medicine, shared a fascinating story that reveals the power of Patient Priorities Care in action.

Wu was a high school student nearly fifty years ago when his father, dying of cancer, asked his son for a solemn promise. *Your mother has dedicated her life to us and after I am gone I am asking you to take care of her.*

Wu has honored that promise for the forty-five years that his mother, Louise Wu, has been a widow. Even though Mrs. Wu lives in Pittsburgh and her son in New Haven, the two are in constant touch. Each evening between 6 and 7 p.m., he calls his mother and they talk on the phone. Four times each year he travels to Pittsburgh to visit with her. "I also check on her remotely with a camera in her living room, kitchen, bedroom, and bathroom to make sure she hasn't fallen or needs help," he explains.

One day, when his mother was in her late eighties, Wu suggested they go through the Patient Priorities Care process.

"Why are we doing this?" she asked.

"I want to honor your wishes," he told her. He began the process convinced that he knew what she most wanted—what her *one thing* was. They worked their way through the Patient Priorities Care website to determine which of four areas generally mattered most to her.

Was it connections to family and friends? To her community or perhaps to spiritual or religious activities?

Was it ensuring the best possible health with effective management of symptoms and discomfort?

242

Was it enjoying life through personal growth, recreation, hobbies?

Or was it functioning with dignity and independence, living and moving independently and safely?

Wu knew what it was. How could he not! He was her son, her devoted caretaker for forty-five years. The answer, he felt sure, was family connections. Quite surprisingly, however, he was wrong.

As Wu and his mother went through the Patient Priorities Care framework with its series of questions and areas for discussions, it became clear.

"When we were discussing function, her ability to get up and move, she talked more and more about that. She was getting very discouraged about not being able to walk, and she has to use a walker now. And particularly the inconvenience of having to get up to go to the bathroom every night so many times. Working through the materials, I learned that functioning, self-sufficiency, dignity, and independence were most important to her." He found that she was embarrassed by urinary incontinence and wanted to sustain her ability to get to the bathroom in time to prevent it.

"It was enlightening to me to learn that it was something different than I thought," Wu told us. "It helped me better understand her in providing better care." About a year later it was clear to Wu that his mother needed help in the home, but she had long been resistant to having a stranger there. But when Wu brought up the subject, he explained to her that "this is going to help [her] get around, help [her] mobility. And with that, she was more accepting and [now] she looks forward to the visiting home aid to help a few times a week."

The next question for Wu and his mother was what to do with this new-found knowledge? How would he apply it in challenging situations? Sometime after Wu and his mother had gone through the Patient Priorities Care process together, she developed a swollen left leg.

"Standard practice would have been to suspect a blood clot, send her to the emergency room, get an ultrasound, and then get started on blood thinning medicines," Wu said. "But in her case, I have to weigh the risks of the treatments and harm, the blood-thinning medicines put you at more risk for bleeding, and particularly bleeding in your head, brain and causing a stroke or cerebral hemorrhage. Just the act of getting her from her home to the emergency room, waiting in the emergency room, waiting for the ultrasound, and factoring in all those things, because she gets tired very easily and she's not going to be sitting in an emergency room doing very well. Also we still have a lot of COVID around and am I putting her more at risk for harm, possibly getting an infection in the hospital. Even with the swelling, she was still able to walk around. I discussed it with her and explained the risks of having a blood clot in her leg going to her lungs and dying or going to the hospital for an ultrasound and possible treatment with blood-thinning medications and risks of bleeding. She said, 'I am still able to walk and let's see how it goes.'"

Fortunately, despite taking the non-standard approach, over the next several days the swelling receded and she remained able to ambulate with a walker at home. In looking back, Wu said, "I think the treatment would've done her more harm."[153]

The simplicity of this approach is appealing. Patients are able to define their priorities in consultation with a member of their care team or they can choose to do it themselves through a detailed, online tool. Tinetti notes that both methods equip patients with a summary that can be shared with their provider team to help guide the decision-making process for care. The summary, as illustrated with Mrs. Wu, defines a health outcome goal for each patient, identifies the health problem of greatest concern to the patient, and zeros in on those treatments the patient is willing and unwilling to take.

Establishing priorities with patients requires physicians, nurses, and others to undergo simple online training sessions focused on getting at the one thing patients want most. Clinicians learn to translate a patient's priorities into health care decisions. The Patient Priorities Care website (patientprioritiescare.org) offers conversation guides to help clinicians speak with patients in different settings, including the emergency department, ambulatory or hospital settings.

Doctors often ridicule new approaches to treating patients. *Enough is enough!* they protest. Enough of new innovations every other month. Physicians are overwhelmed, wary, and understandably so. Tinetti is sensitive to this concern and emphasizes that the patient priorities approach is "not on top of what else you are doing. It is a way of making sure doing what you already do helps patients get what they want from their health care."

Research into the efficacy of Patient Priorities Care has been encouraging. Other researchers writing in the periodical *Geriatrics* in July 2023 concluded that the patient priorities approach to care "can reliably translate 'what matters' into specific, realistic, and actionable health outcome goals. Implementation requires motivated workforce champions, supportive leadership, and uncomplicated EHR tools."

Dr. Christine Ritchie sees important overlap between her work on *care delivery without walls* and Patient Priorities Care "where many of the decisions are preference sensitive because there isn't a clear evidence base for one direction or another," she told us. "You could almost think of it as a checklist for making sure that critical issues of care that often are under-addressed in our current medicalized health care system are actually addressed."

While the Patient Priorities Care approach is simple enough, getting there will be a challenge. Three significant gaps exist within the system that make it more difficult to ensure every patient receives priorities-based care. One issue is a lack of adequately trained staff. Medical education and training traditionally focus on diseases rather than patient preferences. Another challenge is siloed systems of communicating goals across transitions of care, so that physicians and nurses are well aware of the patient's desires. Add to that the opinions of as many as a half dozen specialists caring for an older adult and the complexity of care increases exponentially. Lastly is the corporatization of healthcare that leans on standardized approaches to care that help efficiencies in systems and payment models that are not always accommodating to patient desires.

WHAT WOULD DR. BUTLER THINK?

We suspect that Dr. Butler would warmly embrace Patient Priorities Care, just as we suspect that he would celebrate the progress in caring for older adults since his passing in 2010. Butler was born just a few years shy of a century ago and there is something joyous about reflecting upon his life and work. He saw suffering that many others did not see or, worse, saw and accepted as inevitable to the aging process. He set out to ease that suffering wherever possible, so that older adults might live healthier, more fulfilling lives. Butler's vision has been a guiding light throughout this book and we suspect that he would applaud men and women in these pages for both revealing and mitigating suffering—a monumental achievement.

A final word honoring one of the many gifts Butler passed on to current and future generations: Psychologist William Damon, PhD, a professor of education at Stanford, lauded Butler's pioneering work in devising a therapy called *life review*. As a psychiatrist, Butler was concerned with the problem of increasing depression in aging patients. He believed that the depressive symptoms of his patients stemmed from the aimless way they remembered their pasts. He developed a method for helping people conduct life reviews that find positive benefits in all earlier experiences—even ones that appeared unfortunate at the time.

Butler believed that reflective life reviews would promote "intellectual and personal growth, and wisdom." Among the psychological benefits he noted were: the resolution of old conflicts; an optimistic view of one's future; "a sense of serenity, pride in accomplishment"; a "feeling of having done one's best"; a capacity to enjoy present pleasures such as humor, love, nature, and contemplation; and "a comfortable acceptance of the life cycle, the universe, and the generations." This, of course, is a compelling list of the main pillars of psychological health.[154]

While elements of the life review process can bring on sadness, even grief, Butler writes in *Why Age?* that the "consequences of these steps include expiation of guilt, exorcism of problematic childhood identifications . . . the reconciliation of family relationships, the transmission of knowledge and values to those who follow." Butler honors the desire for legacy—an impulse "so profound in older people to leave something behind when they die. This may be children and grandchildren, lasting work or art, or even memories in the minds of others. . . . bequeathing intellectual and spiritual knowledge."

Ultimately, it means passing along something of themselves, perhaps the example of a life well lived. Not perfectly lived, but a life where the person did his or her best, loved generously, and sought to bring a sense of joy and comfort to anyone they encountered who might have been suffering in some way, physically or emotionally. What a fine legacy, indeed.

ACKNOWLEDGMENTS

The men and women about whom we write in these pages constitute the soul of our book. These are the innovators, the healers who have worked to achieve Dr. Robert Butler's goal of recognizing and mitigating suffering among older adults and who generously shared their stories with us.

We thank the following physicians: Christine Cassel and Diane Meier, pioneers who worked with Dr. Butler, Mt. Sinai, Eric De Jonge and George Taler, MedStar Health and the American Academy of Home Care Medicine, Mary Tinetti and Barry Wu, Yale School of Medicine, Sharon Inouye, Harvard Medical School and editor-in-chief of JAMA Internal Medicine, Bruce Kinosian, the University of Pennsylvania, Thomas H. Lee, Press Ganey Associates and Brigham & Women's Hospital, Boston, Barbara Paris, Maimonides Medical Center in Brooklyn, Tom Edes, the U.S. Department of Veterans Affairs, Christine Ritchie, Harvard's Massachusetts General Hospital, Bruce Leff, Johns Hopkins, Arlene Bierman, the Agency for Healthcare Research and

Quality (AHRQ), Linda V. DeCherrie, Medically Home, Emily Brower, Trinity Health, Michigan, Harold Pincus, Columbia University and Health and Aging Policy Fellowship, Christian Furman, the University of Louisville, David Oliver, the UK National Health Service, Helen Fernandez, Mt. Sinai in New York, and Eric Widera, the University of California, San Francisco, and Kristofer Smith, Chief Medical Officer, Optum at Home.

We thank, as well, Prof. David Meyers, Brown University, James C. Pyles, retired principal Powers Pyles Sutter & Verville, Washington, DC, Leslie Pelton and Dan Schummers, the Institute for Healthcare Improvement, Susan Reinhard of AARP Public Policy Institute, Adam Boehler, Rubicon Founders, Nashville, Rosemary Gibson formerly of the Robert Wood Johnson Foundation, Kimberly Church, the Veterans Administration, Thomas Lally, Bloom Healthcare, Peggy Tighe formerly of the American Academy of Home Care Medicine, Jennie Chin Hansen, former CEO American Geriatrics Society, Margarita Estevez, Syracuse University, Natasha Curry, the UK National Health Service Nuffield Trust, and Stefania Ilinca, the World Health Organization, European Office.

We are grateful to many colleagues and friends here at Northwell Health who helped us along the way including Drs. David Battinelli, Jill Kalman, Peter Silver, Lawrence Smith, Tara Liberman, Santiago Lopez,

ACKNOWLEDGMENTS

Alexander Rimar, Edith Burns, Philip Solomon, David Siskind, Konstantinos "Kostas" Deligiannidis, Joanne Gottridge, John Young, Manish Sapra, Blaine Greenwald, and Isabella Newman. We thank other Northwell colleagues including Merryl Siegel, Irina Mitzner, Sharon Cummings, Tara Pearse, Tara Anglim, Mary Curtis, Susan Kwiatek, Christie Ulbricht, Kathleen Cascio, Brian Aquart, Shivani Rajput, Jeanne Gabriel, Stephanie Ganci, and Brandi Schneider. Also within Northwell Health we are grateful to the teams within the Division of Geriatrics and Palliative Medicine, House Calls, Northwell Health at Home (home care), Hospice Care Network, Department of Psychiatry, and the Division of Geriatric Psychiatry. We are especially grateful to our colleague Terry Lynam for his thoughtful editing of the book at various stages along the way.

We thank the Commonwealth Fund for permission to reprint an informative issue brief on family caregiving by Barbara Lyons and Jane Andrews. We are also grateful to our families for their patience and support.

We are particularly grateful to Terry Fulmer, President of the John A. Hartford Foundation, for her visionary decisions to fund many of the programs we write about that have improved life for older Americans. We owe a special debt to Dr. Kedar Mate, President and CEO of the Institute for Healthcare Improvement, for his thoughtful and moving Preface to the book.

APPENDIX

AUTHOR'S NOTE

The following is an insightful paper published by the New York-based Commonwealth Fund and reprinted here with permission. The paper was written by Barbara Lyons and Jane Andrews. Lyons is an expert on Medicaid and Medicare policy issues and holds a PhD in health policy and finance from Johns Hopkins University Bloomberg School of Public Health. Jane Andrews has held a variety of positions at the Centers for Medicare and Medicaid Services. She holds an M.H.S. in health finance and management from the Johns Hopkins Bloomberg School of Public Health.

COMMONWEALTH FUND ISSUE BRIEF:
POLICY OPTIONS TO SUPPORT FAMILY CAREGIVING FOR MEDICARE BENEFICIARIES AT HOME

By Barbara Lyons and Jane Andrews
November 2023

- **Issue:** Current Medicare policy overlooks the

significant challenges facing tens of millions of family members caring for Medicare beneficiaries at home. Without improved support from Medicare, the need for home-based services will continue to outpace the ability of family caregivers to provide care.

- **Goals:** Identify specific changes to Medicare policies to promote a more equitable system of home-based care that supports beneficiaries and their family caregivers.
- **Methods:** Interviews with twenty aging, disability, and health policy experts, as well as six focus groups and an online survey of one thousand family caregivers.
- **Key Findings:** We identified a range of short- and longer-term policy options to better support family caregivers and improve care for Medicare beneficiaries at home. These policies seek to: 1) increase Medicare coverage of in-home services and supports for family caregivers; 2) provide financial support for family caregivers; 3) expand the availability and accessibility of resources and navigational support for families; and 4) conduct research on ways to advance equity, alleviate caregiver burden, and reduce disparities in access to home-based services.
- **Conclusion:** A wide range of Medicare policy changes could help families care for beneficiaries

and create a more equitable system of home-based care.

INTRODUCTION

The Medicare program, which covers much of the medical and hospital costs for those age sixty-five and older and younger people with disabilities, does not cover many of the costs associated with care at home. Medicare policy overlooks the financial, emotional, and physical pressure on the millions of family members providing home care to beneficiaries who are sick, frail, or limited by disability.[155] They typically work without payment, recognition, or in-home paid care assistance.

The number of family caregivers needed will only increase as the population ages and people can remain at home as a result of advances in technology, such as remote monitoring services. Beneficiaries also have expressed a growing preference to be cared for at home rather than in institutions, amplified by the pandemic.

This policy brief highlights administrative and legislative actions that Medicare could take to better recognize and support family caregivers. We based our recommendations on interviews with twenty aging, disability, and health policy experts who examined ideas suggested in six focus groups and an online survey of one thousand family caregivers of differing genders, ages, races, and ethnicities (see "How We Conducted This Study" for further details on our methods). Highlights from the caregiver survey are summarized in a companion data brief, including that as many as two-thirds of family caregivers feel overwhelmed, anxious, and depressed about their work, and they are often unsure of where to turn for help.

KEY POLICY OPTIONS

Coverage of In-Home Services and Supports for Family Caregivers
1. Eliminate the homebound requirement in the Medicare home health benefit and provide equitable access to existing in-home services.
2. Incentivize Medicare Advantage organizations to offer broader supplemental benefits to beneficiaries and their caregivers, like food and respite care.
3. Increase opportunities for ACOs to provide benefits that support beneficiaries and their caregivers at home.
4. Use the annual Medicare physician fee schedule to increase beneficiary and caregiver services in traditional Medicare, in-home support services, and education and training.
5. Use the CMS Innovation Center, which tests new payment and delivery models, to assess the quality and cost of in-home services to accelerate Medicare's future coverage of these benefits.
6. Expand Medicare coverage of additional benefits for beneficiaries and their caregivers.

COVER IN-HOME SERVICES AND SUPPORTS FOR FAMILY CAREGIVERS

1. *Eliminate the homebound requirement in the Medicare home health benefit and provide equitable access to existing in-home services.* With limited

exceptions, Medicare reimburses in-home health services only if beneficiaries are confined to the home. In the short term, the Centers for Medicare and Medicaid Services (CMS), the federal agency that administers the Medicare program, could promote more equitable access to the home health benefit. For example, CMS could ensure that its prospective payment system for home health does not restrict access to medically complex beneficiaries and underserved populations[156] and to others who are qualified, like those without a prior hospitalization.[157] CMS also could provide more accessible information on benefits, rules, and appeal rights. Further, the agency could encourage testing in the ACO REACH model, a newly redesigned, equity-focused, value-based payment model that permits accountable care organizations (ACOs) to waive the Medicare homebound requirement for access to home health services, and could add similar flexibility to other models.[158]

An interim option for CMS is to offer greater flexibility in considering who is homebound.[159] Over the longer term, Congress could eliminate the homebound requirement and add other improvements to the home health benefit.

2. *Incentivize Medicare Advantage organizations to offer broader supplemental benefits to beneficiaries and their caregivers, like food and respite care,* which allows caregivers a short-term break. Half of

all Medicare beneficiaries are enrolled in Medicare Advantage plans. In the short term, CMS could use regulatory and payment mechanisms to encourage Medicare Advantage organizations to offer more comprehensive supplemental benefits, such as respite and adult day care, and non-primarily health-related benefits like food and nonmedical transportation, to reduce caregiver burden and families' out-of-pocket expenses for home-based care.[160] CMS also could incentivize plans to offer additional benefits that address the nonmedical drivers of health, especially in areas with higher levels of underserved populations.

Over the longer term, Congress could expand eligibility for special supplemental benefits beyond the statutorily defined chronically ill population. This expansion would make more beneficiaries eligible for nonmedical benefits like meals and groceries, which may positively affect health outcomes.[161] Expanded eligibility could include those with low-income subsidy eligibility who can currently receive nonmedical supplemental benefits in plans offered under the Medicare Advantage Value-Based Insurance Design Model.[162]

3. *Increase opportunities for ACOs to provide benefits that support beneficiaries and their caregivers at home.* Through the current ACO REACH Model, policymakers can learn from testing whether care

management home visits prevent hospitalizations or whether waiving the Medicare homebound requirement improves access to home health services.

Additionally, the 2023 physician fee schedule offers incentives for new ACOs in the Medicare Shared Savings Program, one of CMS's original value-based alternative payment models, to receive funds to address the social needs of people with Medicare that could help underserved beneficiaries and caregivers at home.[163] Some Medicare ACOs already offer services that help caregivers manage beneficiaries' care like case management, health and functional assessments, mobile technologies, and 24/7 clinical support. Over the longer term, the annual physician fee schedule could encourage more ACOs to provide other services that support beneficiaries and caregivers at home.

4. *Use the annual Medicare physician fee schedule to increase beneficiary and caregiver services in traditional Medicare, in-home support services, and education and training.* An example of a recent service, added by the 2023 fee schedule, is coverage of some dental services because they are inextricably linked to the clinical success of an otherwise covered medical service.[164] Accordingly, CMS could propose other services that help beneficiaries and caregivers at home.

Use of currently available codes for providing

services, like assessments of caregiver health risk and their ability to care for a beneficiary with dementia, could be increased through greater provider education and awareness. These caregiver-specific services may not be sufficiently publicized in existing provider resources such as the Medicare Learning Network or tip sheets.[165] New codes finalized for 2024, including the code to provide payment when practitioners train and involve caregivers to support patients with certain diseases or illnesses (e.g., dementia) in carrying out a treatment plan,[166] require provider outreach and education to ensure their use. In addition, agencies also could ensure caregivers receive the required training. For example, home health agency conditions of participation require caregiver preparedness training.[167]

5. *Use the CMS Innovation Center, which tests new payment and delivery models, to assess the quality and cost of in-home services to accelerate Medicare's future coverage of these benefits.* For example, the CMS Innovation Center has recently developed the GUIDE Model, a dementia care model that tests patient- and family-centered care, including the needs of caregivers, over a period of years.[168] The model includes respite care but could include other supports for family caregivers. Caregiver supports could be tested more broadly beyond dementia, in other models as well.

6. *Expand Medicare coverage of additional benefits for beneficiaries and their caregivers.* Traditional Medicare provides little to no coverage of stand-alone benefits like home health aides, personal care services, adult day care, mental health services for caregivers, and home safety modifications like grab bars. Health benefits for beneficiaries like vision, dental, and hearing that could help ease the burden on caregivers are also generally not covered, nor are nonmedical benefits like meals and transportation[169] that help support home-based care. One short-term strategy is to ensure that community-based organizations are fully funded and their social benefits are accessible to all Medicare beneficiaries. Over the longer term, expanding such benefits in traditional Medicare would require congressional action.

PROVIDE FINANCIAL SUPPORT FOR FAMILY CAREGIVERS

Financial Support for Family Caregivers

1. Test payment options to compensate family caregivers for services provided in the home.
2. Provide financial assistance for paid help and/or reimburse caregivers' out-of-pocket expenses paid on behalf of the beneficiary, like durable medical equipment, transportation costs, and housing assistance.

Interviews and survey responses revealed that caregivers experience significant financial difficulties as a result of their work, often jeopardizing their own financial security to support Medicare beneficiaries at home.[170] The following policy recommendations are aimed at easing some of this financial pressure.

1. *Test payment options to compensate family caregivers for services provided in the home.* Providing compensation or a stipend has precedent in other federal programs, including Medicaid and Veterans Affairs (VA) health care.[171] Currently, some Medicare Advantage plans offer family caregivers limited reimbursement for the hours of care provided at home. Over the short term, CMS could encourage or incentivize these plans to offer better caregiver reimbursement. The CMS Innovation Center also could test compensation or a stipend under a Medicare model. Over the longer term, including caregiver compensation in traditional Medicare would require congressional legislation.

2. *Provide financial assistance for paid help and/or reimburse caregivers' out-of-pocket expenses paid on behalf of the beneficiary, like durable medical equipment, transportation costs, and housing assistance.* Over the short term, Medicare Advantage plans can provide broader special supplemental benefits including help with cost sharing for covered services as long as they fall under the limited Special Supplemental Benefits for the Chronically

Ill statutory definition. As noted elsewhere, without a legislative change to broaden eligibility for these nonmedical benefits, they are not available to all Medicare Advantage enrollees. Over the long term, financial assistance or reimbursement to caregivers would require congressional legislation.

MAKE INFORMATIONAL RESOURCES AND NAVIGATIONAL SUPPORT AVAILABLE AND ACCESSIBLE

Availability and Accessibility of Informational Resources and Navigational Support

1. Create an office within CMS to identify, train, and prepare culturally sensitive and accessible health care advocates.
2. Incorporate easy-to-understand and relevant information about caregiving benefits in Medicare publications and other resources for beneficiaries and providers.
3. Ensure that Medicare Advantage plans provide enrollees with information on supplemental benefit offerings each year and that enrollees understand how to access and use benefits.
4. Establish a centralized resource that links caregivers to other benefits programs (e.g., SNAP).
5. Include caregivers in the development of beneficiaries' care plans and identify the caregiver in the medical record.

Survey findings suggest that most family caregivers spend significant time looking for information on covered benefits and services under their family members' health insurance plans. Similarly, some focus group participants said they struggled to find the information they needed. The following policies could make information more readily accessible to caregivers.

1. *Create an office within CMS to identify, train, and prepare culturally sensitive and accessible health care advocates.* These advocates could help beneficiaries and caregivers access Medicare and community services, avoid unnecessary out-of-pocket payments for covered services, and coordinate care between Medicare and community-based organizations. The final 2024 physician fee schedule responds in part to this need by including a principal illness navigation service that would help patients with cancer or debilitating illness and their caregivers identify and access providers. The final rule also includes a community health integration service that would refer families to community health workers who could provide services to address unmet social needs.[172]

2. *Incorporate easy-to-understand and relevant information about caregiving benefits into Medicare publications and other resources for beneficiaries and providers.* Over the short term, CMS could modify existing print and online publications to include caregiving benefits and training information using the VA's Program of General Caregiver Support Services (which provides peer mentoring to caregivers of veterans) as a model.[173] CMS also could update the Medicare Beneficiary Ombudsman webpage to

recognize caregivers and provide more helpful links to resources for caregivers.[174]

3. *Ensure that Medicare Advantage plans provide enrollees with information on supplemental benefit offerings each year and that enrollees understand how to access and use benefits.* In the short term, CMS could enforce requirements that plans inform beneficiaries of available Medicare Advantage supplemental benefits, including those that impact caregivers, during open enrollment and during the coverage period. CMS also could ensure that Medicare Advantage plans promptly communicate all benefits information—including social benefits such as meals, in-home supports, and caregiver respite and compensation—to beneficiaries and their caregivers. In addition, MA plans could ensure enrollee understanding of the benefit and its delivery by tracking utilization and performing outreach to enrollees where uptake is low.

4. *Establish a centralized resource that links caregivers to other benefits programs.* Other programs that can help support Medicare beneficiaries and their caregivers include Medicaid, the Supplemental Nutrition Assistance Program, the Medicare Savings Program, and the Low-Income Subsidy program. The Administration for Community Living's MIPPA program provides grants to states that support information on programs that help to save on Medicare

costs, particularly for low-income beneficiaries.[175] This resource could serve as an example for establishing a centralized resource hub for caregivers.

5. *Include caregivers in the development of beneficiaries' care plans and identify the caregiver in the medical record.* Over the short term, CMS could enforce existing rules to ensure caregivers are included in care transitions across all care settings. Similarly, the White House's executive order on caregiving directs the HHS Secretary to ensure that hospitals actively include family caregivers in the discharge planning process.[176] Additionally, the Caregiver Advise, Record, Enable (CARE) Act model legislation, enacted in forty-two states, the District of Columbia, Puerto Rico, and the U.S. Virgin Islands, requires hospitals to record the name of the family caregiver on the medical record; inform the family caregiver when their loved one is to be discharged; and provide the family caregiver with education and instruction of the medical tasks to be performed for the patient at home.[177] Finally, the 2022 Recognize, Assist, Include, Support, and Engage (RAISE) Act Strategy points out that medical records increasingly include documentation of names, support, abilities, and preferences of caregivers, so caregivers will be included on the beneficiary's care teams.[178]

APPENDIX

PROMOTE RESEARCH TO ADVANCE KNOWLEDGE ABOUT CAREGIVER BURDEN AND ADDRESS DISPARITIES

Research to Advance Knowledge About Caregiver Burden and Address Disparities

1. Collect quality, patient satisfaction, and outcomes data to analyze caregivers' burden.
2. Make access to telehealth services and other technologies more equitable.

---

1. *Collect quality, patient satisfaction, and outcomes data to analyze caregivers' burden.* CMS could increase use of existing data or collect more caregiver data through the National Health and Aging Trends Study and the National Study of Caregiving and other databases to better recognize caregivers and their needs.[179] Consistent with the CMS Framework for Health Equity 2022–2032, CMS could collect and compare data on the availability and use of home-based services and differences in quality outcomes between populations, including underserved populations.[180] The CMS Innovation Center also could require its model participants to collect beneficiary and caregiver data to access to home-based services, differences in quality, and caregiver burden.

As CMS develops broad quality measures to support its goals, policymakers could determine if existing performance measures for Medicare home-based services are adequate.[181] Agencies also could incorporate caregiver questions into all Medicare patient satisfaction surveys, like the Consumer Assessment of Healthcare Providers and Systems.[182]

2. *Make access to telehealth services and other technologies more equitable.* Telehealth supports beneficiaries' care at home and can reduce the burden on caregivers. For example, telehealth may provide an alternative to transporting a beneficiary to the doctor's office. As use of telehealth and "hospital at home" services continue to increase post-pandemic, data on their impacts across underserved populations will become available.[183] CMS could use these data to identify gaps in access by income, geography, and underrepresented populations.[184]

DISCUSSION

We found that policymakers have a range of short- and long-range options to support family caregivers providing care for Medicare beneficiaries at home. Many options do not require congressional action to better recognize caregivers and pay for services that support them. The Centers for Medicare and Medicaid Services, for example, could increase the availability and uptake of the Medicare home health benefit, assuring that the full range of services is provided. The agency could use the annual physician fee schedule, regulatory and payment mechanisms for Medicare Advantage and ACOs, and the Innovation Center's authority to accelerate Medicare's future coverage of benefits and services to support beneficiaries and their family caregivers. And CMS could provide resources to help Medicare beneficiaries and their caregivers navigate medical and support needs at home and access home-based services, including emerging technologies. In providing home-based services and caregiver supports, CMS also could build on the efforts and experiences of other programs, including Medicaid, Veterans Affairs, and the Administration for Community Living.

Beyond these short-term efforts, Congress could add a home-based benefit and take legislative action to help ensure equity in services for all Medicare beneficiaries. Such actions could help recognize the pivotal role of family caregivers and reduce the significant burdens they face.

HOW WE CONDUCTED THIS STUDY

To identify policy options, we drew on accumulated evidence and new research findings conducted for this project between January and March 2023.

New data on the direct experiences and views of family caregivers were collected by PerryUndem, in collaboration with the Center for Advancing Innovative Policy. Family caregivers (including relatives, partners, neighbors, friends, and other individuals) were defined as adults age eighteen and older providing unpaid assistance with personal care and household manage-ment tasks for a Medicare beneficiary (a person age six-ty-five and older or a younger person with disabilities) currently or in the past four years. An online survey of one thousand current and recent caregivers was con-ducted, drawing on six online focus group discussions. Differences by race, ethnicity, gender, income, and geog-raphy were analyzed.

We identified policy options through an environmen-tal scan and twenty structured interviews with Medicare experts, government officials, providers, and community representatives. These options build on the views and priorities of family caregivers of Medicare beneficiaries identified through the focus groups and national survey.

ENDNOTES

CHAPTER 1

1　Robert N. Butler, *Why Survive? Being Old in America* (Johns Hopkins University Press, 1975).

2　Ibid.

3　Robin Marantz Henig, "Why Survive?" *New York Times*, Sept. 14, 1975.

4　This eventually led to commercial production of lovastatin by Merck (1970s). Commercial laboratories including AstraZeneca (Propranolol), Squibb (Lisinopril), and Merck (Lipitor) produced the first major commercial medications to treat both high blood pressure and cholesterol. Propranolol and lisinopril, also used to manage blood pressure, are among the most widely used drugs in the world, as is lovastatin for cholesterol management. These drugs were approved for commercial use in 1987.

　　Akiro Endo. "A Historical Perspective on the Discovery of Statins". *Proc Jpn Acad Ser B Phys Biol Sci.* 2010;86(5):484–93. doi: 10.2183/pjab.86.484. PMID: 20467214; PMCID: PMC3108295.

5　Matthew D. Ritchey MD, Hillary K. Wall, Mary G. George, Janet S. Wright. "US Trends in Premature HDisease Mortality over the Past 50 Years: Where do we go from here?" *Trends Cardiovasc Med.* 2020 Aug;30(6):364–374. doi: 10.1016/j.tcm.2019.09.005. Epub 2019 Sep 27. PMID: 31607635; PMCID: PMC7098848.

6　Rebecca L Siegel, Kimberly D Miller, Nikita Sandeep Wagle, Ahmedin Jemal, "Cancer statistics, 2023." *CA Cancer J Clin.* 2023 Jan;73(1):17–48. doi: 10.3322/caac.21763. PMID: 36633525.

7　Ibid.

8　Dr. Amit Arora, Sally Greenbrook, "The Geriatric Medicine Workforce 2022". *British Geriatrics Society,* September 1,

2022. https://www.bgs.org.uk/GMworkforce22, Accessed on September 1, 2023.

9 Robert N. Butler, *Why Survive: Being Old in America* (Johns Hopkins University Press, 1975).

10 Mark Mather, et al., "America's Changing Population: What to Expect in the 2020 Census," *Population Bulletin* 74, no. 1 (2019).

11 Wendy Fox-Grage, "The Growing Racial and Ethnic Diversity of Older Adults". *AARP Public Policy Institute,* April 18, 2016. https://blog.aarp.org/thinking-policy/the-growing-racial-and -ethnic-diversity-of-older-adults

12 Center for Health Workforce Studies School of Public Health, University at Albany, *The Impact of the Aging Population on the Health Workforce in the United States: Summary of Key Findings* (March 2006), pp 1–19. https://www.chwsny.org/wp -content/uploads/2015/09/ImpactofAging2005_Excerpt.pdf, accessed November 1, 2023.

13 Fuchs, Matt. "Virtual Reality Could Give Relief to Seniors." *New York Times*, May 6, 2022. https://www.nytimes.com/2022/05/06 /well/mind/virtual-reality-therapy-seniors.html.

14 Martha Lally and Suzanne Valentine-French, *Lifespan Development: A Psychological Perspective, Second Edition* (Open Textbook Library, 2022).

15 Emily M. Mitchell, PhD, "Concentration of Healthcare Expenditures and Selected Characteristics of Persons with High Expenses, U.S. Civilian Noninstitutionalized Population, (*Agency for Healthcare Research and Quality Statistical Brief,* #540 2019, February 2022).

16 Douglas Martin, "Robert Butler, Aging Expert, Is Dead at 83" *New York Times*, July 7, 2010.

17 W. Andrew Achenbaum, *Robert N. Butler, MD Visionary of Healthy Aging* (Oxford University Press, 2013).

18 Gregg A. Warshaw, Elizabeth J. Bragg, "The Training of Geriatricians in the United States: Three Decades of Progress" *Journal of the American Geriatrics Society*, August 2003.

19 Graphic Detail Daily Chart "America's Missing Doctors. Where are all the Geriatricians?" *The Economist*, September 22, 2023.

20 Good News Network, "More than 70% of Older Americans Feel Younger Than They Actually Are—And Are Embracing Aging" (Nov. 20, 2021) https://www.goodnewsnetwork.org/70-percent-of -older-americans-feel-younger/.

21 Helene H. Fung, PhD, "Aging in Culture" *The Gerontologist* Vol. 53, No. 3, 369–377) doi:10.1093/geront/gnt024.

22 It is often suggested that the United States ranks well below many

nations in the quality of health care and perhaps, by certain metrics, that may be the case. Certainly, the idea that the U.S. does not measure up to other nations in health care delivery has become an article of faith in the media in the past decade plus. Reports often rank the U.S. at the bottom among ten or so wealthy nations. In 2021, the three top rankings went to Norway (population 5.4 million), the Netherlands (population 18 million), and Australia (population 26 million). These are tiny countries with relatively homogeneous populations. The population of the United States, for example, is nearly 70 times that of Norway, 20 times larger than the Netherlands, and 13 times the size of Australia. These are not ideal apples to apples comparisons. And certainly, in the U.S., unhealthy lifestyles are rampant, far less so in those three nations, while in the U.S. we face an epidemic of gun violence. These reflect poorly on U.S. culture, certainly, but they are not areas where doctors, hospitals and other providers can be found at fault. The U.S. suffers, as well, from continuing underinvestment in primary care, public health, and social services. The Commonwealth Fund also reports that US adults are least likely to have longstanding relationship with a primary care provider, lower number of physician visits and number of practitioners, more individuals with multiple chronic conditions. This underinvestment can impact health outcomes, health education, and trust in health care. Eric C. Schneider, Arnav Shah, Michelle M. Doty, Roosa Tikkanen, Katharine Fields, Reginald D. Williams II, "Mirror, Mirror 2021 — Reflecting Poorly: Health Care in the U.S. Compared to Other High-Income Countries" (*Commonwealth Fund*, Aug. 2021). https://doi.org/10.26099/01dv-h208.

CHAPTER 2

23 American Academy of Home Care Medicine. During the early years of the American Academy of Home Care Medicine the core group making the case for increased reimbursements for home visits included Dr. Taler as well as Connie Row, CEO of the group, Drs. Peter Boling from Richmond, Virginia, Gresham Bayne from San Diego.

24 De Jonge's work promoting home visits earned him a reputation as a leader in the home-based field and in 2003, he was honored as the National House Call Physician of the Year by the American Academy of Home Care Medicine.

25 The physicians on the policy team at the Academy included George Taler at MedStar Health in Washington DC, Peter Boling at Virginia Commonwealth University in Richmond, VA, Jim

Pyles, an attorney, and Gresham Bayne from Call-Doc, a company in San Diego, CA. Joining the core team a bit later were De Jonge, Drs. Bruce Kinosian at the University of Pennsylvania, Bruce Leff at Johns Hopkins, and Thomas Edes at the Department of Veterans Affairs, along with Bob Sowislo, a CFO from a home-based care program called the Visiting Physicians Association.

26 Emily M. Mitchell, PhD, "Concentration of Healthcare Expenditures and Selected Characteristics of Persons with High Expenses, U.S. Civilian Noninstitutionalized Population" (*Agency for Healthcare Research and Quality Statistical Brief #540 2019, February 2022*).

27 Oseroff BH, Ankuda CK, Bollens-Lund E, Garrido MM, Ornstein KA, "Patterns of Healthcare Utilization and Spending Among Homebound Older Adults in the USA: an Observational Study", *Journal of General Internal Med.* 2023.

28 Prior to enrolling in the home-care VA patients were costing the agency $46,000 per patient per year (1996). However, after these VA patients enrolled in the home-care program the total per person cost of care declined to just $17,000 per year including the $6,000 cost for the home-based primary care program, a decline of nearly $30,000 per patient per year. The explanation for why this happened was simple: Patients in the home-based program were so well cared for that they rarely needed to go to the emergency room and even less frequently had to be admitted for hospital stays thus avoiding some of the most expensive medical services.

29 Usually for meetings with members and staff on Capitol Hill, Pyles was accompanied by Taler and Edes, but they were frequently joined by De Jonge, Boling, and later, Bruce Kinosian.

30 De Jonge told us: "We did conservative calculations. If you expanded this across the country and you reduced Medicare costs by just 10 percent, which is the minimum shown by our preliminary data. There was VA data and other program data we can point to locally. If you save 10 percent or $5,000 per patient per year and multiplied that by two million people (5 percent of Medicare fee-for service population), and do the math, that came to $10 billion in savings per year."

31 K. Eric De Jonge, Namirah Jamshed, Daniel Gilden, Joanna Kubisiak, Stephanie R Bruce, George Taler, "Effects of Home-Based Primary Care on Medicare Costs in High-Risk Elders" (*Journal of the American Geriatrics Society*, 18 July 2014) [The study focused on team-based house call care for 722 frail persons over five years in the MedStar Washington Hospital Center

House Calls Program. De Jonge and his team aimed to "provide or coordinate all needed primary care and arrange specialty care that is compatible with beneficiaries' values and preferences. This avoids the phenomenon of discordant care that can result when multiple specialists care for an individual with complex needs without coordination by a primary care team."

32 DeJonge KE, Taler G, Boling PA. "Independence at home: community-based care for older adults with severe chronic illness" *Clin Geriatr Med.* 2009 Feb;25(1):155–69, ix. doi: 10.1016/j.cger .2008.11.004. PMID: 19217500.

33 Daniela J. Lamas, MD, "Admitted to Your Bedroom: Some Hospitals Try Treating Patients at Home," *(New York Times,* April 27, 2015).

34 Christine Ritchie and Bruce Leff, "Home-Based Care Reimagined: A Full-Fledged Health Care Delivery Ecosystem Without Walls" *Health Affairs,* May 2022.

CHAPTER 3

35 Sharon K. Inouye et al. "Geriatric Syndromes: Clinical, Research, and Policy Implications of a Core Geriatric Concept." *Journal of the American Geriatrics Society* 55, no. 5 (May 2007): 780–91. https://doi.org/10.1111/j.1532–5415.2007.01156.x).

36 Mary E. Tinetti, DI Baker, G McAvay, EB Caus, P Garrett, M Gottschalk, ML Koch, K Trainor, RI Horwitz. "A Multifactorial Intervention to Reduce the Risk of Falling among Elderly People Living in the Community" *N Engl J Med* 1994; 331: 821–827.

37 Inouye SK, Bogardus ST Jr, Charpentier PA, Leo-Summers L, Acampora D, Holford TR, Cooney LM Jr. "A Multicomponent Intervention to Prevent Delirium in Hospitalized Older Patients." *N Engl J Med.* 1999 Mar 4;340(9):669–76 doi: 10.1056/NEJM1999 03043400901. PMID: 10053175.

38 "What Is Delirium?," *American Delirium Society,* accessed November 26, 2023, https://americandeliriumsociety.org/patients -families/what-is-delirium/.

39 Tammy T. Hshieh, et al. "Hospital Elder Life Program: Systematic Review and Meta-analysis of Effectiveness" *Am J Geriatr Psychiatry.* 2018 Oct;26(10):1015–1033. doi: 10.1016/j.jagp.2018.06.007. Epub 2018 Jun 26. PMID: 30076080; PMCID: PMC6362826.

40 Tinetti ME, Williams TF, Mayewski R. "Fall Risk Index for Elderly Patients based on Number of Chronic Disabilities" *Am J Med.* 1986 Mar;80(3):429–34. doi: 10.1016/0002–9343(86)90717–5. PMID: 3953620.

41 Tinetti ME, Speechley M, Ginter SF. "Risk factors for Falls among

Elderly Persons Living in the Community" *N Engl J Med* 1988 Dec 29;319(26):1701–7. doi: 10.1056/NEJM198812293192604. PMID: 3205267.

42   Mary E. Tinetti, DI Baker, G McAvay, EB Claus, P Garrett, M Gottschalk, ML Koch, K Trainor, RI Horwitz, et al. "A Multifactorial Intervention to Reduce the Risk of Falling among Elderly People Living in the Community." *N Engl J Med* 331, no. 13 (September 29, 1994): 821–27. https://doi.org/10.1056/nejm199409293311301.

43   Ibid.

44   Reuters. "Study Offers Tips for Elderly on Reducing Risk of Falling" (*New York Times*), September 30, 1994 Section A, Page 17. https://www.nytimes.com/1994/09/30/us/study-offers-tips-for-elderly-on-reducing-risk-of-falling.html, accessed December 1, 2023.

45   Mary E. Tinetti, Dorothy I. Baker, Mary King, Margaret Gottschalk, Terence E. Murphy, Denise Acampora, Bradley P. Carlin, Linda Leo-Summers, Heather Allure. "Effect of Dissemination of Evidence in Reducing Injuries from Falls," *New Engl J Med* (359, no. 3 (July 17, 2008): 252–61, https://doi.org/10.1056/nejmoa0801748.)

46   Ibid.

47   Ohn Mar, "Program Reduces Falls by Elderly, Study Finds," *New York Times*, August 11, 2008, https://www.nytimes.com/2008/08/12/health/12fall.html..

48   Fried TR, McGraw S, Agostini JV, Tinetti ME. "Views of Older Persons with Multiple Morbidities on Competing Outcomes and Clinical Decision-making." *J Am Geriatr Soc.* 2008 Oct;56(10):1839–44. doi: 10.1111/j.1532–5415.2008.01923.x. Epub 2008 Sep 2. PMID: 18771453; PMCID: PMC2596278.

49   Marcum ZA, Gellad WF. "Medication Adherence to Multidrug Regimens." *Clin Geriatr Med* 2012 May;28(2):287–300. doi: 10.1016/j.cger.2012.01.008. PMID: 22500544; PMCID: PMC3335752.

50   Sharon Inouye, Sidney T. Bogardus, Peter A Charpentier, Linda Leo-Summers, Denise Acampora, Theoford Holford, Leo M.Cooney. "A Multicomponent Intervention to Prevent Delirium in Hospitalized Older Patients." *N Engl J Med.* 1999 Mar 4;340(9):669–76. doi: 10.1056/NEJM199903043400901. PMID: 10053175.

51   Tammy T Hshieh, Tinghan Yang, Sarah L Gartaganis, Jirong Yue, Sharon Inouye. "Hospital Elder Life Program: Systematic Review and Meta-analysis of Effectiveness." *Am J Geriatr Psychiatry.*

2018 Oct;26(10):1015–1033. doi: 10.1016/j.jagp.2018.06.007. Epub 2018 Jun 26. PMID: 30076080; PMCID: PMC6362826.

52 Tammy T. Hshieh, Jirong Yue, Esther Oh, Margaret Puelle, Sarah Dowal, Thomas Travison, Sharon K. Inouye. "Effectiveness of Multicomponent Nonpharmacological Delirium Interventions." *JAMA Internal Medicine* 175, no. 4 (April 1, 2015): 512. https://doi.org/10.1001/jamainternmed.2014.7779.

53 Tammy T. Hshieh, Tinghan Yang, Sarah L. Gartaganis, Jirong Yue, Sharon Inouye. "Hospital Elder Life Program: Systematic Review and Meta-analysis of Effectiveness." *Am J Geriatr Psychiatry.* 2018 Oct;26(10):1015–1033. doi: 10.1016/j.jagp.2018.06.007. Epub 2018 Jun 26. PMID: 30076080; PMCID: PMC6362826

54 "AGS CoCare—Help," AGS CoCare, *American Geriatrics Society*, accessed December 1, 2023, https://help.agscocare.org/

55 "AGS CoCare: HELP," AGS CoCare, *American Geriatrics Society*, accessed November 1, 2023, https://help.agscocare.org /About_AGS_CoCare_program_help.

56 "AGS CoCare: HELP & Age-Friendly," AGS CoCare, *American Geriatrics Society*, accessed November 1, 2023, https://help .agscocare.org/ags_cocare_help_and_age_friendly.

57 Andrea Macdonald, "Dr. Inouye: Delirium an under Diagnosed Area of Brain Aging and Degeneration," *ideaXme*, June 11, 2020, https://radioideaxme.com/2020/03/06/dr-inouye-delirium -an-under-diagnosed-area-of-brain-aging-and-degeneration/.

58 Fick DM, Auerbach AD, Avidan MS, Busby-Whitehead J, Ely EW, Jones RN, Marcantonio ER, Needham DM, Pandharipande P, Robinson TN, Schmitt EM, Travison TG, Inouye SK. "Network for Investigation of Delirium Across the U.S. (NIDUS): Advancing the Field of Delirium with a New Interdisciplinary Research Network" *J Gerontol Nurs.* 2017 May 1;43(5):4–6. doi: 10.3928/00989134 –20170411-01. PMID: 28448673.

59 Eva M. Schmitt EM, Jane S. Saczynski JS, Cyrus M. Kosar, Jones RN, David C. Alsop, Fong TG, Erin Metzger, Zara Cooper, Marcantonio ER, Thomas Travison, Sharon K. Inouye; SAGES Study Group. "The Successful Aging after Elective Surgery (SAGES) Study: Cohort Description and Data Quality Procedures." *J Am Geriatr Soc* 2015 Dec; 63(12):2463–2471. doi: 10.1111/jgs.13793. Epub 2015 Dec 14. PMID: 26662213; PMCID: PMC4688907.

CHAPTER 4

60 Steven R. Counsell, Christopher M. Callahan, Amna B. Buttar, Daniel O. Clark, Kathryn I. Frank. "Geriatric Resources for

Assessment and Care of Elders (GRACE): a New Model of Primary Care for Low-Income Seniors" *J Am Geriatr Soc* 2006 Jul;54(7):1136–41.) doi: 10.1111/j.1532–5415.2006.00791.x. PMID: 16866688.

61 Fulmer T, Mezey M, Bottrell M, Abraham I, Sazant J, Grossman S, Grisham E. "Nurses Improving Care for Healthsystem Elders (NICHE): Using Outcomes and Benchmarks for Evidenced-based Practice." *Geriatr Nurs* 2002 May-Jun;23(3):121–7. doi: 10.1067/mgn.2002.125423. PMID: 12075275.

62 Michael J. Barry, Susan Edgman-Levitan. "Shared Decision Making--Pinnacle of Patient-centered Care" *N Engl J Med.* 2012 Mar 1;366(9):780–1. Doi: 10.1056/NEJMp1109283. PMID: 22375967.

63 Ibid.

64 Ibid.

65 Leslie Pelton, "Age-Friendly Health Systems," *Institute for Health-care Improvement*, May 13, 2019, https://www.ihi.org /sites/default/files/2023–09/IHI_Business_Case_for_Becoming _Age_Friendly_Health_System.pdf.

66 *Northwell Health Advanced Illness Collaborative: Advanced Illness Definition*, adapted from Institute for Healthcare Improvement and The Coalition to Transform Advanced Care 2014–2015.

67 Jette DU, Stilphen M, Ranganathan VK, Passek SD, Frost FS, Jette AM. Validity of the AM-PAC "6-Clicks" inpatient daily activity and basic mobility short forms. *Phys Ther.* 2014 Mar;94(3):379–91. doi: 10.2522/ptj.20130199. Epub 2013 Nov 14. PMID: 24231229.

68 Julia R Adler-Milstein, et al. "Health System Approaches and Experiences implementing the 4Ms: Insights from 3 Early Adopter Health" *Gerontological Society of America*, 29 May, 2023, pp 2627–2639.

69 Timothy W. Farrell, "Fighting Ageism with Age-Friendly Care," *University of Utah Health*, May 25, 2023, https://uofuhealth .utah.edu/notes/2022/10/age-friendly-health-system.

70 Ibid.

CHAPTER 5

71 Michael J. Denham. "Lord Amulree (1900–83): The Indefatigable Advocate of Older Persons" *Journal of Medical Biography.* 2006;14(4):236–242. doi:10.1177/096777200601400412

72 Michael J. Denham. "Dr Marjory Warren CBE MRCS LRCP (1897–1960): the Mother of British Geriatric Medicine," *Journal*

of Medical Biography. 2011;19(3):105–110. doi:10.1258/jmb
.2010.010030.

73 Caroline Richmond, "Dame Cicely Saunders, Founder of the
Modern Hospice Movement, Dies," The BMJ, December 28, 2023,
https://www.bmj.com/content/suppl/2005/07/18/331.7509.DC1.

74 Michael J. Denham. "Lord Amulree (1900–83): The Indefatigable
Advocate of Older Persons." Journal of Medical Biography.
2006;14(4):236–242. doi:10.1177/096777200601400412.

75 Michael J. Denham. "Dr Marjory Warren CBE MRCS LRCP
(1897–1960): the Mother of British Geriatric Medicine." Journal
of Medical Biography. 2011;19(3):105–110. doi:10.1258/jmb
.2010.010030.

76 Caroline Richmond, "Dame Cicely Saunders, Founder of the
Modern Hospice Movement, Dies," The BMJ, December 28, 2023,
https://www.bmj.com/content/suppl/2005/07/18/331.7509.DC1.

77 Keith Schneider, "Dr. Jack Kevorkian Dies at 83; a Doctor Who
Helped End Lives," New York Times, June 3, 2011, https://www
.nytimes.com/2011/06/04/us/04kevorkian.html.

78 Alfred F. Connors et al. "A Controlled Trial to Improve Care
for Seriously Ill Hospitalized Patients: The Study to Understand
Prognoses and Preferences for Outcomes and Risks of Treatments
(SUPPORT)" JAMA. 1995;274(20):1591–1598. doi:10.1001
/jama.1995.03530200027032.

79 Kenneth E. Covinsky, JD Fuller K Yaffe, CB Johnston, MB
Hamel, J Lynn, JM Teno, RS Phillips. "Communication
and Decision-Making in Seriously Ill Patients: Findings of
the SUPPORT Project"; JAGS 48:S187-S193,2000. doi.
org/10.1111/j.1532–5415.2000.tb03131.x

80 Education in Palliative and End-of-Life Care: EPEC, Feinberg
School of Medicine, Core Curriculum, accessed November 24, 2023,
https://www.bioethics.northwestern.edu/programs/epec
/curricula/core-curriculum.html.

81 Maria T. Carney, Diane E. Meier. "Palliative care and end-of-
life issues" Anesthesiol Clin North Am. 2000 Mar;18(1):183–
209DOI: 10.1016/s0889–8537(05)70156–5.

82 "On Our Own Terms: Moyers on Dying: Shows," BillMoyers.
com, January 8, 2024, https://billmoyers.com/series/on-our-own
-terms-moyers-on-dying/.

83 Bill Moyers, Living With Dying, BillMoyers.com, 2015, https:
//billmoyers.com/content/on-our-own-terms-living-with-dying
/#:~:text=No%20one%20wants%20to%20think,their%20
families%20and%20their%20culture.

84 Julie Salamon, "That Mystery That No One Wants to Think

about Is Getting More Complex," *New York Times*, September 8, 2000, https://www.nytimes.com/2000/09/08/movies/tv-weekend -that-mystery-that-no-one-wants-to-think-about-is-getting -more-complex.html.

85 "ACGME Program Requirements for Graduate Medical Education in Geriatric . . .," ACGME Program Requirements for Graduate Medical Education in Geriatric Medicine, June 12, 2022, https: //www.acgme.org/globalassets/pfassets/programrequirements /125_geriatricmedicine_2022.pdf, pp22–23.

86 Diane Meier. Palliative Care Pioneer Dr. Diane Meier on How People Struggle with Serious, Sometimes Terminal, Illness, Interview by Amy Goodman with *Democracy Now*, April 8, 2010, *Mount Sinai*, https://www.mountsinai.org/about/newsroom/2010 /palliative-care-pioneer-dr-diane-meier-on-how-people-struggle -with-serious-sometimes-terminal-illness.

87 Barbara Sadick, "Straight Talk about Palliative Care: What Everyone Should Know," *The Wall Street Journal*, September 14, 2014, https://www.wsj.com/articles/straight-talk-about-palliative -care-what-everyone-should-know-1410724839.

88 Renske Visser, Erica Borgstrom, Richard Holti. "The Overlap Between Geriatric Medicine and Palliative Care: A Scoping Literature Review" *J Appl Gerontol*. 2021 Apr;40(4):355–364.

89 Ibid.

90 Ibid.

91 Keith Schneider, "Dr. Jack Kevorkian Dies at 83; a Doctor Who Helped End Lives," *New York Times*, June 3, 2011, https://www .nytimes.com/2011/06/04/us/04kevorkian.html.

CHAPTER 6

92 "S535 (1R)—An Act concerning pronouncement of death and amending P.L. 1983, c.308.," accessed November 27, 2023, https: //www.nj.gov/lps/dcj/codification/S%20535%201R.pdf.

93 *National Survey on Long-Term Care: Expectations and Reality* (The Associated Press - NORC Center for Public Affairs Research, May 19, 2014), Associated Press, https://www.norc.org/research /library/national-survey-on-long-term-care--expectations-and -reality.html.

94 Susan C. Reinhard et al. "Valuing the Invaluable: 2023 Update." *AARP Public Policy Institute*. March 8, 2023.pp.1–32. https://doi .org/10.26419/ppi.00082.006.

95 American Psychological Association, *Mental and Physical Health Effects of Family Caregiving*, 2015, https://www.apa.org/pi /about/publications/caregivers/faq/health-effects.

ENDNOTES

96 Richard Schulz and Scott R Beach. "Caregiving as a risk factor for mortality: The Caregiver Health Effects Study" *JAMA* 282, 1999.2215–2219. doi:10.1001/jama.282.23.2215.

97 National Opinion Research Center. *Long term care in America: Expectations and realities.* Chicago, IL: The Associated Press and NORC; 2014, https://www.longtermcarepoll.org/wp-content /uploads/2017/11/AP-NORC-Long-term-Care-2014_Trend _Report.pdf, accessed October 30, 2023.

98 AARP, "Timeline: AARP at 65," *AARP*, November 10, 2023, https: //www.aarp.org/politics-society/history/info-2023/aarp-at-65. html.

99 Susan C. Reinhard et al. "Valuing the Invaluable: 2023 Update." *AARP Public Policy Institute*. March 8, 2023. pp.1–32. https: //doi.org/10.26419/ppi.00082.006.

100 Susan C. Reinhard, "Valuing the Invaluable: 2015 Update," *The Lund Report*, July 17, 2015, https://www.thelundreport.org /content/valuing-invaluable-2015-update.

101 Susan C. Reinhard, Carol Levine, and Sarah Samis, "Home Alone: Family Caregivers Providing Complex Chronic Care", *AARP Policy Institute*, October 2012, https://media.uhfnyc.org/filer _public/c5/a6/c5a6c494–42ac-42dd-be0c-99d802e2c985/home _alone_new_final_102612.pdf.

102 American Psychological Association, ed., "As Americans Age, Caregiving Challenges Increase: Six Questions for Caregiving Expert Stephen Zarit, PhD" *American Psychological Association*, 2014, https://www.apa.org/news/press/releases/2014/06/caregiving -challenges.

103 Richard Schulz and Scott R. Beach. "Caregiving as a Risk Factor for Mortality: The Caregiver Health Effects Study", *JAMA* 282, 1999.2215–2219. doi:10.1001/jama.282.23.2215).

104 The RAISE Act Family Caregiving Advisory Council and The Advisory Council to Support Grandparents Raising Grandchildren, "National Strategy to Support Family Caregivers", *Administration for Community Living*. September 21, 2022, https://acl.gov/sites /default/files/RAISE_SGRG/NatlStrategyToSupportFamily Caregivers.pdf.

105 Susan C. Reinhard, et al. "Valuing the Invaluable: 2023 Update." *AARP Public Policy Institute*. March 8, 2023.pp.1–32. https: //doi.org/10.26419/ppi.00082.006.

106 "The Program of Comprehensive Assistance for Family Caregivers," Veterans Affairs, February 22, 2023, https://www.va.gov/family -member-benefits/comprehensive-assistance-for-family-caregivers/.

107 Joan M. Griffin, Brystana G. Kaufman, Lauren Bangerter, Diane E. Holland, Catherine Vanderboom, Cory Ingram, Ellen M. Wild, Ann Marie Dose, Carole Stiles, Virginia H. Thompson. "Improving Transitions in Care for Patients and Family Caregivers Living in Rural and Underserved Areas: The Caregiver Advise, Record, Enable (CARE) Act." *J Aging Soc Policy.* 2022 Feb 13:1–8. Doi: 10.1080/08959420.2022.2029272. Epub ahead of print. PMID: 35156557.

108 *FACT SHEET: Vice President Harris Announces That American Rescue Plan Investments in Home and Community-Based Care Services for Millions of Seniors and Americans with Disabilities Reach About $37 Billion Across All 50 States* (The White House, December 11, 2023), The White House, https://www.whitehouse.gov/briefing-room/statements-releases/2023/12/11/fact-sheet-vice-president-harris-announces-that-american-rescue-plan-investments-in-home-and-community-based-care-services-for-millions-of-seniors-and-americans-with-disabilities-reach-about-37/.

CHAPTER 7

109 Gilfillan also served as Chief Medical Officer for Independence Blue Cross and General Manager of their AmeriHealth New Jersey subsidiary, as SVP of National Contracting at Coventry Health, and as head of Geisinger insurance companies.

110 Steve Strongwater, MD, *Atrius Health History* (Atrius Health website, 2023).

111 Sachin H. Jain, MD, "6 Habits of High Value Health Care Organizations," *Forbes*, April 13, 2016.

112 Charles Kenney, "The Best Practice: How the New Quality Movement Is Transforming Medicine" *PublicAffairs* 2008.

113 Cara S. Lesser, Paul B. Ginsburg, and Kelly J. Devers, "The End of an Era: What Became of the Managed Care Revolution 2001?" *NIH Health Serv Research*, Feb. 2003.

114 Donald M. Berwick, Thomas W. Nolan, John Whittington. "The Triple Aim: Care, Health, and Cost." *Health Aff* (Millwood). 2008 May-Jun;27(3):759–69. Doi: 10.1377/hlthaff.27.3.759. PMID: 18474969.

115 Among others playing an important role at the time were researchers at Dartmouth College, including Drs. Elliott S. Fisher, Jack Wennberg and others at the Dartmouth Atlas of Health Care Project, which documented "glaring variations in how medical resources are distributed and used in the United States." The Dartmouth research showed that health care spending was

devouring an ever-greater share of the nation's gross domestic product (GDP), but with "little evidence that the amount we are spending is producing better health outcomes for populations or individual patients . . ." The Dartmouth researchers found that far too much of the volume of care produced no benefit for patients. In other words, the fee-for-service system was rife with waste. Identifying "waste and eliminating it would not only help provide financial security for the Medicare program without loss of value to the people it covers, it would also help finance the expansion of coverage." And in a particularly demanding indictment, Dartmouth researchers found that supply of medical services often drove demand rather than the other way around. See FAQ—Dartmouth Atlas of Health Care.

116 Gilfillan noted: "The MSSP program was based on the Accountable Care Organizations (ACOs) model described by Elliott Fisher and colleagues. The basic idea was to create an organization that included physicians, hospitals and other providers acting together to take responsibility for the cost and quality outcomes for population, much as Kaiser, HCHP and Geisinger Group Practices had done. It was a test to see if these integrated provider based Accountable Care Organizations could duplicate Group Practice outcomes for the Traditional Medicare population."

117 In addition to Atrius Health, original Pioneers included the likes of Bellin Health in Green Bay, WI, Dartmouth-Hitchcock Medical Center in Hanover, NH, Michigan Pioneer ACO, and Montefiore Medical Center in New York.

118 Richard Barasch, Chairman and CEO Universal American Corp (Are Medicare ACO's Working? Experts Disagree, KFF Health News, Oct. 21, 2015).

119 National Association of ACOs (2020 Medicare ACO Program Results; saved $4.1 billion over an eight-year period, more than any other ACO program.

120 While Gilfillan, Berwick, and Conway were the leaders, there were others who played critically important roles along with them including Jon Blum, Deputy Administrator and Director of the Center for Medicare, Liz Richter, Deputy Director Center for Medicare, John Pilotte, Director of the Performance Based Payment Policy Group, Terri Postma, MD Chief Medical Officer Performance Based Payment Policy Group, Mai Pham, MD Seamless Care Models Group/Chief Innovation Officer CMMI, Sean Cavanaugh, Seamless Care Models Group/Deputy Director CMMI.

CHAPTER 8

121 Angelica Peebles, "Medicare's Worst-Paying Specialty Is Luring Billions From Wall Street," *Bloomberg*, (February 10, 2022).

122 Fyodor Dostoevsky, *The Brothers Karamazov* (1879).

123 Arnold S. Relman, M.D., "The New Medical-Industrial Complex," *New Engl J Med* Oct. 23, 1980.

124 Relman A. "Arnold Relman—the Last Angry Doctor" Interview by Dennis L. Breo. *JAMA.* 1991 Jun 5;265(21):2864–5, 2869. doi: 10.1001/jama.265.21864. PMID: 2033748.

125 Robert Lowes, "Arnold Relman, Medicine's Long-time Conscience, Dies at 91" *Medscape Medical News,* (June 19, 2014).

126 Bryan Marquard, "Dr. Arnold Relman, 91; ex-N.E. Journal of Medicine editor" *Boston Globe,* (June 20, 2014).

127 Nancy Ochieng, Jeannie Fuglesten Biniek, Meredith Freed, Anthony Damico, and Tricia Neuman, "Medicare Advantage in 2023: Enrollment Update and Key Trends" *KFF Health News,* (Aug. 9, 2023).

128 Interview with Professor David Meyers, PhD, Brown University.

129 CompaniesMarketCap and Macrotrends.net.

130 Richard Gilfillan, Donald M. Berwick, "Medicare Advantage, Direct Contracting, And The Medicare 'Money Machine,' Part 1: The Risk-Score Game" *Health Affairs* (Sept. 29, 2021).

131 Gilfillan, Berwick, "Medicare Advantage, Direct Contracting, And The Medicare 'Money Machine,' Part 2: Building On The ACO Model" *Health Affairs* (Sept. 30, 2021)

132 Ibid.

133 Reed Abelson and Margot Sanger-Katz, "The Cash Monster Was Insatiable: How Insurers Exploited Medicare for Billions" *New York Times* (Oct. 8, 2022).

134 WNCT/AP/WNCN, Sept. 25, 2019. In June of 2019 Conway was arrested for drunk driving while his two young daughters were passengers in the vehicle. In October of 2019, Conway was convicted of driving while impaired and two counts of misdemeanor child abuse for having his two young daughters in the car.

135 Gilfillan and Berwick, "Born on Third Base: Medicare Advantage Thrives on Subsidies, Not Better Care", *Health Affairs* (March 27, 2023).

136 David J. Meyers, PhD, MPH; Amal N. Trivedi, MD, "Trends in the Source of New Enrollees to Medicare Advantage From 2012 to 2019" *JAMA Health Forum,* (Aug. 12, 2022).

CHAPTER 9

137 Maddy Reinert, Danielle Fritze, & Theresa Nguyen. "The State

of Mental Health in America 2023: Key Findings" Mental Health America, page 9, Alexandria VA, (October 2022). https://mhanational.org/sites/default/files/2023-State-of-Mental-Health-in-America-Report.pdf.

138 Amy Fiske, Julie Loebach Wetherell, Margaret Gatz. "Depression in Older Adults," *Annu Rev Clin Psychol.* 2009;5:363–89. doi: 10.1146/annurev.clinpsy.032408.153621. PMID: 19327033; PMCID: PMC2852580.

139 Interview with Dr. Blaine Greenwald Vice Chairman of Psychiatry and Geriatric Psychiatry at Northwell.

140 Oliver Whang, "Physician Burnout Has Reached Distressing Levels, New Research Finds" *New York Times*, (Sept. 29, 2022).

141 National Academies of Sciences, Engineering, and Medicine, National Academy of Medicine, and Committee on Systems Approaches to Improve Patient Care by Supporting Clinician Well-Being, "Taking Action Against Clinician Burnout: A Systems Approach to Professional Well-Being," *National Academy of Sciences Press*, (October 23, 2019), https://doi.org/10.17226/25521.

142 Colin P. West, MD, PhD; Liselotte N. Dyrbye, MD, MHPE; Christine Sinsky, MD, Mickey Trockel, MD, PhD; Michael Tutty, PhD; Laurence Nedelec, PhD; Lindsey E. Carlasare, MBA; Tait D. Shanafelt, MD, "Resilience and Burnout Among Physicians and the General US Working Population" *JAMA Network Open*, (July 2, 2020).

143 Ibid.

144 Agency for Healthcare Research and Quality, "*Physician Burnout*" Accessed November 1, 2023. https://www.ahrq.gov/prevention/clinician/ahrq-works/burnout/index.html.

145 National Academies of Sciences, Engineering, and Medicine; Health and Medicine Division; Board on Health Care Services; Committee on Implementing High-Quality Primary Care. "Implementing High-Quality Primary Care: Rebuilding the Foundation of Health Care" Robinson SK, Meisnere M, Phillips RL Jr, McCauley L, editors. Washington (DC): National Academies Press (US); 2021 May 4. PMID: 34251766.

146 Robin M. Weinick, Maria Torroella Carney, Trisha Milnes, Arlene S. Bierman, in partnership with AHRQ. "Optimizing Health and Function as We Age Roundtable Report" AHRQ Publication No. 23–0059, (July 2023).

147 Penn Leonard Davis Institute for Health Economics Seminar, *Reforming Primary Care for a 21st Century Health Care System*, Rachel Werner, Ishani Ganguli, Linda McCauley, J. Nwando

Olayiwola, Ellen-Marie Whelan, Patrick Conway by Hoag Levins. https://ldi.upenn.edu/our-work/research-updates/how-can-we-fix-primary-care/, accessed on November 26, 2023.

148 W. Andrew Achenbaum, *Robert N. Butler, MD Visionary of Healthy Aging* (Oxford University Press, 2013).

CHAPTER 10

149 CBS, *60 Minutes : Dr. Farmer's Remedy for World Health.* (May 1, 2008). https://www.cbsnews.com/news/dr-farmers-remedy-for-world-health/, accessed December 1, 2023.

150 Kidder, Tracy. Mountains beyond Mountains. New York :Random House, (2003).

151 Ellen Barry and Alex Traub. Paul Farmer, "Pioneer of Global Health, Dies at 62," *The New York Times,* (Feb 22, 2022).

152 Paul Farmer: "Accompaniment as Policy" *Harvard Magazine,* (May 25, 2011).

153 Crystal Gwizdala, *Caregiving from 460 Miles Away: A Geriatrician's Experience Caring for His Mother*, published by Yale School of Medicine Nov. 20, 2023 [some of this from Patient Priority Care Retreat, January 18, 2024].

154 William Damon, PhD, "Purpose and the Life Review" *Psychology Today*, July 28, 2021.

APPENDIX

155 Barbara Lyons and Jane Andrews, "Caring for Medicare Beneficiaries at Home: Experience and Priorities of Family Caregivers" Commonwealth Fund, Oct. 2023.

156 Centers for Medicare and Medicaid Services, "CMS Framework for Health Equity 2022–2032," (*CMS*, Apr. 2022).

157 Bipartisan Policy Center, *Optimizing the Medicare Home Health Benefit to Improve Outcomes and Reduce Disparities* (BPC, Apr. 2022).

158 Centers for Medicare and Medicaid Services, "ACO REACH," *CMS*, Aug. 18, 2023.

159 Centers for Medicare and Medicaid Services, "*Home Health Agencies: CMS Flexibilities to Fight COVID-19*" CMS, May 10, 2023.

160 Kathryn A. Coleman, "Reinterpretation of 'Primarily Health Related' for Supplemental Benefits," letter to Medicare Advantage Organizations and Section 1876 Cost Contract Plans, Medicare Drug and Health Plan Contract Administration Group, *Centers for Medicare and Medicaid Services*, Apr. 27, 2018.

161 Amelia Whitman, et al., *Addressing Social Determinants of*

ENDNOTES

Health: Examples of Successful Evidence-Based Strategies and Current Federal Efforts (U.S. Department of Health and Human Services, Assistant Secretary of Planning and Evaluation, Apr. 2022).

162 Centers for Medicare and Medicaid Services, "Medicare Advantage Value-Based Insurance Design Model," June 23, 2023.

163 Centers for Medicare and Medicaid Services, "Medicare and Medicaid Programs; CY 2023 Payment Policies Under the Physician Fee Schedule and Other Changes to Part B Payment and Coverage Policies," *Federal Register* 87 (Nov. 18, 2022): 69404.

164 CMS, "Medicare and Medicaid Programs," 2022.

165 Centers for Medicare and Medicaid Services, "Tip Sheet for Providers: Caregiving Education," n.d.

166 Centers for Medicare and Medicaid Services, "Calendar Year (CY) 2024 Medicare Physician Fee Schedule Final Rule," CMS fact sheet, Nov. 2, 2023.

167 BPC, *Optimizing Medicare Home Health*, 2022.

168 Centers for Medicare and Medicaid Services, "Guiding an Improved Dementia Experience (GUIDE) Model," n.d.

169 Whitman, et al., *Addressing Social Determinants*, 2022.

170 Lyons and Andrews, *Caring for Medicare Beneficiaries*, 2023.

171 Salom Teshale, Wendy Fox-Grage, and Kitty Purington, *Paying Family Caregivers Through Medicaid Consumer-Directed Programs: State Opportunities and Innovations* (Administration for Community Living, National Academy for State Health Policy, and the John A. Hartford Foundation, Apr. 2021); and U.S. Department of Veterans Affairs, "VA Caregiver Support Program," Aug. 21, 2023.

172 CMS, "Calendar Year (CY) 2024 Final Rule," 2023.

173 Veterans Affairs, "VA Caregiver Support," 2023.

174 Centers for Medicare and Medicaid Services, "Medicare Beneficiary Ombudsman (MBO)," last updated Sept. 6, 2023.

175 Administration for Community Living, "Medicare Improvements for Patients and Providers (MIPPA)," last updated Oct. 3, 2022.

176 The White House, "Executive Order on Increasing Access to High-Quality Care and Supporting Caregivers," Apr. 18, 2023.

177 AARP, "State Law to Help Family Caregivers," n.d.

178 The Recognize, Assist, Include, Support, and Engage (RAISE) Act Family Caregiving Advisory Council and the Advisory Council to Support Grandparents Raising Grandchildren, *2022 National Strategy to Support Family Caregivers* (RAISE Act Family

Caregiving Advisory Council and the Advisory Council to SGRG, Sept. 21, 2022).

179 National Health and Aging Trends Study, "National Study of Caregiving (NSOC)," n.d.

180 CMS, *Framework for Health Equity*, 2022.

181 Douglas B. Jacobs, et al., "Aligning Quality Measures Across CMS —The Universal Foundation," *New England Journal of Medicine* 388, no. 9 (Mar. 2, 2023): 776–79.

182 Centers for Medicare and Medicaid Services, "Consumer Assessment of Healthcare Providers & Systems (CAHPS)," last updated Sept. 6, 2023.

183 Centers for Medicare and Medicaid Services, "CMS Waivers, Flexibilities, and the Transition Forward from the COVID-19 Public Health Emergency," fact sheet, Feb. 27, 2023.

184 Emily Gillen, Robin Duddy-Tenbrunsel, and Gabriel Miller, "Limited Internet Access in Underserved Communities Could Drive Disparities in Telehealth Utilization.

INDEX

INDEX